In the name of God, Most Gracious, Most Merciful

About The Author

The author, who writes under the pen-name HARUN YAHYA, was born in Ankara in 1956. Having completed his primary and secondary education in Ankara, he then studied arts at Istanbul's Mimar Sinan University and philosophy at Istanbul University. Since the 1980s, the author has published many books on political, faith-related and scientific issues. Harun Yahya is well-known as an author who has written very important works disclosing the imposture of evolutionists, the invalidity of their claims and the dark liaisons between Darwinism and bloody ideologies such as fascism and communism.

His pen-name is made up of the names "Harun" (Aaron) and "Yahya" (John), in memory of the two esteemed prophets who fought against lack of faith. The Prophet's seal on the cover of the books is symbolic and is linked to the their contents. It represents the Qur'an (the final scripture) and the Prophet Muhammad, the last of the prophets. Under the guidance of the Qur'an and sunnah, the author makes it his purpose to disprove each one of the fundamental tenets of godless ideologies and to have the "last word", so as to completely silence the objections raised against religion. The seal of the final Prophet, who attained ultimate wisdom and moral perfection, is used as a sign of his intention of saying this last word. All author's works center around one goal: to convey the Qur'an's message to people, encourage them to think about basic faith-related issues (such as the existence of God, His unity and the Hereafter), and to expose the feeble foundations and perverted ideologies of godless systems.

Harun Yahya enjoys a wide readership in many countries, from India to America, England to Indonesia, Poland to Bosnia, and Spain to Brazil. Some of his books are available in English, French, German, Spanish, Italian, Portuguese, Urdu, Arabic, Albanian, Russian, Serbo-Croat (Bosnian), Polish, Malay, Uygur Turkish, and Indonesian, and they are enjoyed by readers worldwide.

Greatly appreciated all around the world, these works have been instrumental in many people recovering their faith in God and in many others gaining a deeper insight into their faith. The wisdom, and the sincere and easy-to-understand style gives these books a distinct touch which directly effects any one who reads or studies them. Immune to objections, these works are characterized by their features of rapid effectiveness, definite results and irrefutability. It is unlikely that those who read these books and give serious thought to them can any longer sincerely advocate the materialistic philosophy, atheism or any other perverted ideology or philosophy. Even if they continue to do so, it will be only a sentimental insistence since these books refuted such ideologies from their very foundations. All contemporary movements of denial are now ideologically defeated, thanks to the collection of books written by Harun Yahya.

There is no doubt that these features result from the wisdom and lucidity of the Qur'an. The author modestly intends to serve as a means in humanity's search for God's right path. No material gain is sought in the publication of these works.

Considering these facts, those who encourage people to read these books, which open the "eyes" of the heart and guide them to become more devoted servants of God, render an invaluable service.

Meanwhile, it would just be a waste of time and energy to propagate other books which create confusion in peoples' minds, lead man into ideological chaos, and which, clearly have no strong and precise effects in removing the doubts in peoples' hearts, as also verified from previous experience. It is apparent that it is impossible for books devised to emphasize the author's literary power rather than the noble goal of saving people from loss of faith, to have such a great effect. Those who doubt this can readily see that the sole aim of Harun Yahya's books is to overcome disbelief and to disseminate the moral values of the Qur'an. The success and impact of this service are manifest in readers' conviction.

One point should be kept in mind: The main reason for the continuing cruelty, conflict, and all the ordeals the majority of people undergo is the ideological prevalence of disbelief. This state can only be ended with the ideological defeat of disbelief and by conveying the wonders of creation and Qur'anic morality so that people can live by it. Considering the state of the world today, which leads people into the downward spiral of violence, corruption and conflict, it is clear that this service has to be provided more speedily and effectively. Otherwise, it may be too late.

It is no exaggeration to say that the collection of books by Harun Yahya have assumed this leading role. By the will of God, these books will be a means through which people in the 21st century will attain the peace, justice and happiness promised in the Qur'an.

TO THE READER

In all the books by the author, faith-related issues are explained in the light of Qur'anic verses, and people are invited to learn God's words and to live by them. All the subjects that concern God's verses are explained in such a way as to leave no room for doubt or question marks in the reader's mind. The sincere, plain and fluent style employed ensures that everyone of every age and from every social group can easily understand the books. This effective and lucid narrative makes it possible to read them in a single sitting. Even those who rigorously reject spirituality are influenced by the facts recounted in these books and cannot refute the truthfulness of their contents.

This book and all the other works by Harun Yahya can be read individually or discussed in a group. Those readers who are willing to profit from the books will find discussion very useful in that they will be able to relate their own reflections and experiences to one another.

In addition, it is a great service to the religion to contribute to the presentation and circulation of these books, which are written solely for the good pleasure of God. All the books of the author are extremely convincing, so, for those who want to communicate the religion to other people, one of the most effective methods is to encourage them to read these books.

It is hoped that the reader will take time to look through the review of other books on the final pages of the book, and appreciate the rich source of material on faith-related issues, which are very useful and a pleasure to read.

In them, one will not find, as in some other books, the personal views of the author, explanations based on dubious sources, styles unobservant of the respect and reverence due to sacred subjects, or hopeless, doubt-creating, and pessimistic accounts that create deviations in the heart.

THE
EVOLUTION
DECEIT

Copyright© Harun Yahya 1420 AH / 1999CE

First published by Vural Yayincilik, Istanbul, Turkey in April 1997

First English Edition published in July 1999

Published by:
Ta-Ha Publishers Ltd.
1 Wynne Road
London SW9 OBB
United Kingdom

Website: http://www.taha.co.uk
E-mail: sales@taha.co.uk

All rights reserved. No part of this publication may be reproduced, stored in any retrieval system or transmitted in any form or by any methods, electronic, mechanical, photocopying, recording or otherwise without the prior permission of the publishers.

By Harun Yahya

Translated by: Mustapha Ahmad

A Catalog Record of this book is available from the British Library

ISBN 1-897940-97-1

Printed and bound by:
Kelebek Matbaacilik

www.harunyahya.com
www.evolutiondeceit.com

THE
EVOLUTION DECEIT

The Scientific Collapse of Darwinism and Its Ideological Background

HARUN YAHYA

Ta-Ha Publishers Ltd.
1 Wynne Road London SW9 OBB
United Kingdom

CONTENTS

SPECIAL PREFACE: The Real Ideological Root of Terrorism:
Darwinism and Materialism 14
Introduction 14
The Darwinist Lie: 'Life is conflict'..14
Darwin's Source of Inspiration: Malthus's Theory of Ruthlessness...........15
What 'The Law of the Jungle' Led to: Fascism 16
The Bloody Alliance: Darwinism and Communism....................17
Darwinism and Terrorism 19
ISLAM IS NOT THE SOURCE OF TERRORISM, BUT ITS SOLUTION 20
Islam is a Religion of Peace and Well-Being 20
God Condemns Mischief 21
Islam Defends Tolerance and Freedom of Speech 21
God Has Made the Killing of Innocent People Unlawful 22
God Commands Believers to be Compassionate and Merciful.................23
God Has Commanded Tolerance and Forgiveness 23
Conclusion 24

PART I
THE REFUTATION OF DARWINISM

INTRODUCTION: Why the Theory of Evolution?......................27

FOREWORD: The Greatest Miracle of Our Times: Belief in the
Evolution Deceit 30

CHAPTER 1: To Be Freed FromPrejudice 34
Blind Materialism 36
Mass Evolutionist Indoctrination 38

CHAPTER 2: A Brief History of the Theory 41
Darwin's Imagination 42
Darwin's Racism 44
The Desperate Efforts of Neo-Darwinism...................................45

Trial and Error: Punctuated Equilibrium 46
The Primitive Level of Science in Darwin's Time 47

CHAPTER 3: Imaginary Mechanisms of Evolution 50
Natural Selection 50
"Industrial Melanism" 51
Can Natural Selection Explain Complexity? 53
Mutations 53

CHAPTER 4: The Fossil Record Refutes Evolution 58
Life Emerged on Earth Suddenly and in Complex Forms 61
Molecular Comparisons Deepen Evolution's Cambrian Impasse 64

CHAPTER 5: Tale of Transition from Water to Land 65
Why Transition From Water to Land is Impossible 67

CHAPTER 6: Origin of Birds and Mammals 69
Another Alleged Transitional Form: *Archæopteryx* 70
Speculations of Evolutionists: The Teeth and Claws of *Archæopteryx* 72
Archæopteryx and Other Ancient Bird Fossils 73
The Design of the Bird Feathers 74
The Imaginary Bird-Dinosaur Link 76
What is the Origin of Flies? 77
The Origin of Mammals 78
The Myth of Horse Evolution 80

CHAPTER 7: Deceptive Fossil Interpretations 81

CHAPTER 8: Evolution Forgeries 84
Piltdown Man: An Orangutan Jaw and a Human Skull! 84
Nebraska Man: A Pig's Tooth 86
Ota Benga: The African In The Cage 87

CHAPTER 9: The Scenario of Human Evolution 89
   The Imaginary Family Tree of Man..90
   Australopithecus: An Ape Species...92
   Homo Habilis: The Ape that was Presented as Human 93
   Homo Rudolfensis: The Face Wrongly Joined 98
   Homo Erectus and Thereafter: Human Beings 100
   Homo Erectus: An Ancient Human Race..101
   Neanderthals 104
   Homo Sapiens Archaic, *Homo Heilderbergensis* and Cro-Magnon Man 105
   Species Living in the Same Age as Their Ancestors................................107
   The Secret History of *Homo Sapiens*..108
   A Hut 1.7 Million Years Old 110
   Footprints of Modern Man, 3.6 Million Years Old! 110
   The Bipedalism Impasse of Evolution...112
   Evolution: An Unscientific Faith 113

CHAPTER 10: The Molecular Impasse of Evolution............................116
   The Tale of the "Cell Produced by Chance" 116
   Confessions from Evolutionists...118
   The Miracle in the Cell and the End of Evolution 119
   Proteins Challenge Chance 121
   Left-handed Proteins 125
   Correct Bond is Vital...127
   Zero Probability..127
   Is There a Trial and Error Mechanism in Nature? 128
   The Probability of a Protein Being Formed by Chance is Zero.............129
   The Evolutionary Fuss About the Origin of Life..131
   Miller's Experiment 133
   Miller's Experiment was Nothing but Make-believe 133
   Latest Evolutionist Sources Dispute Miller's Experiment 135
   Primordial World Atmosphere and Proteins...137
   Protein Synthesis is not Possible in Water...138

Another Desperate Effort: Fox's Experiment 138
Inanimate Matter Cannot Generate Life 140
The Miraculous Molecule: DNA 141
Can DNA Come into Being by Chance? 142
Another Evolutionist Vain Attempt: "The RNA World" 145
Confessions From Evolutionists 146
Life is a Concept Beyond Mere Heaps of Molecules 148

CHAPTER 11: Thermodynamics Falsifies Evolution 150
The Myth of the "Open System" 152
The Myth of the "Self Organization of Matter" 153

CHAPTER 12: Design and Coincidence 152
Darwinian Formula! 158
Technology In The Eye and The Ear 159
The Theory of Evolution is the Most Potent Spell in the World 162

CHAPTER 13: Evolutionist Claims and the Facts 165
Variations and Species 165
Antibiotic Resistance and DDT Immunity
are not Evidence for Evolution 167
The Fallacy of Vestigial Organs 170
The Myth of Homology 172
Similar Organs in Entirely Different Living Species 173
The Genetic and Embryological Impasse of Homology 174
Invalidity of the Claim of Molecular Homology 175
The Myth of Embryological Recapitulation 178

CHAPTER 14: The Theory of Evolution: A Materialistic Liability 180
Materialist Confessions 182
The Death of Materialism 185
Materialists, False Religion and True Religion 186

CHAPTER 15: Media: Fertile Ground for Evolution 188
    Wrapped-up Lies 190

CHAPTER 16: Conclusion: Evolution Is a Deceit 192
    The Theory of Evolution has Collapsed 192
    Evolution Can Not Be Verified in the Future Either 193
    The Biggest Obstacle to Evolution: Soul 193
    God Creates According to His Will 194

CHAPTER 17: The Fact of Creation 196
    Honey Bees and the Architectural Wonders of Honeycombs 196
    Amazing Architects: Termites 198
    The Woodpecker 198
    The Sonar System of Bats 199
    Whales 199
    The Design in The Gnat 200
    Hunting Birds with Keen Eyesight 201
    The Thread of the Spider 201
    Hibernating Animals 202
    Electrical Fish 203
    An Intelligent Plan on Animals: Camouflage 203
    Cuttlefish 204
    Different Vision Systems 205
    Special Freezing System 206
    Albatrosses 207
    An Arduous Migration 207
    Koalas 208
    Hunting Ability in Constant Position 209
    The Design In Bird Feathers 209
    Basilisk: The Expert of Walking on Water 210
    Photosynthesis 211

PART II
THE REFUTATION OF MATERIALISM

CHAPTER 18: The Real Essence of Matter 215
    The World Of Electrical Signals 216
    How Do We See, Hear, And Taste? 217
    "The External World" Inside Our Brain 222
    Is The Existence Of The "External World" Indispensable? 225
    Who Is The Perceiver? 226
    The Real Absolute Being 228
    Everything That You Possess Is Intrinsically Illusory 231
    Logical Deficiencies Of The Materialists 235
    The Example Of Dreams 236
    The Example Of Connecting The Nerves In Parallel 237
    The Formation Of Perceptions In The Brain Is Not Philosophy But Scientific Fact 239
    The Great Fear Of The Materialists 240
    Materialists Have Fallen Into The Biggest Trap In History 243
    Conclusion 246

CHAPTER 19: Relativity of Time and the Reality of Fate 249
    The Perception Of Time 249
    The Scientific Explanation Of Timelessness 250
    Relativity In The Qur'an 254
    Destiny 256
    The Worry Of The Materialists 258
    The Gain Of Believers 259

CHAPTER 20: SRF Conferences:
   Activities for Informing the Public About Evolution 261

NOTES 268

SPECIAL PREFACE

The Real Ideological Root of Terrorism: Darwinism and Materialism

Introduction

Most people think the theory of evolution was first proposed by Charles Darwin, and rests on scientific evidence, observations and experiments. However, the truth is that Darwin was not its originator, and neither does the theory rest on scientific proof. The theory consists of an adaptation to nature of the ancient dogma of materialist philosophy. Although it is not backed up by scientific discoveries, the theory is blindly supported in the name of materialist philosophy.

This fanaticism has resulted in all kinds of disasters. Together with the spread of Darwinism and the materialist philosophy it supports, the answer to the question "What is a human being?" has changed. People who used to answer: "God creates human beings and they have to live according to the beautiful morality He teaches", have now begun to think that "Man came into being by chance, and is an animal who developed by means of the fight for survival." There is a heavy price to pay for this great deception. Violent ideologies such as racism, fascism and communism, and many other barbaric world views based on conflict have all drawn strength from this deception.

This article will examine the disaster Darwinism has visited on the world and reveal its connection with terrorism, one of the most important global problems of our time.

The Darwinist Lie: 'Life is conflict'

Darwin set out with one basic premise when developing his theory: **"The development of living things depends on the fight for survival. The strong win the struggle. The weak are condemned to defeat and oblivion."**

According to Darwin, there is a ruthless struggle for survival and an eternal conflict in nature. The strong always overcome the weak, and this enables development to take place. The subtitle he gave to his book *The Origin of Species*, "*The Origin of Species by Means of Natural Selection or the Preservation of Favoured Races in the Struggle for Life*", encapsulates that view.

Furthermore, Darwin proposed that the 'fight for survival' also applied between human racial groups. According to that fantastical claim, 'favoured races' were victorious in the struggle. Favoured races, in Darwin's view, were white Europeans. African or Asian races had lagged behind in the struggle for survival. Darwin went further, and suggested that these races would soon lose the "struggle for survival" entirely, and thus disappear:

> At some future period, not very distant as measured by centuries, the civilised races of man will almost certainly exterminate, and replace the savage races throughout the world. At the same time the anthropomorphous apes... will no doubt be exterminated. The break between man and his nearest allies will then be wider, for it will intervene between man in a more civilised state, as we may hope, even than the Caucasian, and some ape **as low as a baboon, instead of as now between the negro or Australian and the gorilla.**[1]

The Indian anthropologist Lalita Vidyarthi explains how Darwin's theory of evolution imposed racism on the social sciences:

> His (Darwin's) theory of the survival of the fittest was warmly welcomed by the social scientists of the day, and they believed mankind had achieved various levels of evolution culminating in the white man's civilization. By the second half of the nineteenth century racism was accepted as fact by the vast majority of Western scientists.[2]

Darwin's Source of Inspiration: Malthus's Theory of Ruthlessness

Darwin's source of inspiration on this subject was the British economist Thomas Malthus's book *An Essay on the Principle of Population*. Left to their own devices, Malthus calculated that the human population increased rapidly. In his view, the main influences that kept populations under control were disasters such as war, famine and disease. In short, according to this brutal claim, some people had to die for others to live. Existence came to mean "permanent war."

In the 19th century, Malthus's ideas were widely accepted. European upper class intellectuals in particular supported his cruel ideas. In the article **"The Scientific Background of the Nazi 'Race Purification' Programme"**, the importance 19th century Europe attached to Malthus's views on population is described in this way:

> In the opening half of the nineteenth century, throughout Europe, members of the ruling classes gathered to discuss the newly discovered "Population problem" and to devise ways of implementing the Malthusian mandate, to increase the mortality rate of the poor: **"Instead of recommending cleanliness to the poor, we should encourage contrary habits. In our towns we should make the streets narrower, crowd more people into the houses, and court the return of the plague. In the country we should build our villages near stagnant pools, and particularly encourage settlements in all marshy and unwholesome situations,"** and so forth and so on.[3]

As a result of this cruel policy, the weak, and those who lost the struggle for survival would be eliminated, and as a result the rapid rise in population would be balanced out. This so-called "oppression of the poor" policy was actually carried out in 19th century Britain. An industrial order was set up in which children of eight and nine were made to work sixteen hours a day in the coal mines and thousands died from the terrible conditions. The "struggle for survival" demanded by Malthus's theory led to millions of Britons leading lives full of suffering.

Influenced by these ideas, Darwin applied this concept of conflict to all of nature, and proposed that the strong and the fittest emerged victorious from this war of existence. Moreover, he claimed that the so-called struggle for survival was a justified and unchangeable law of nature. On the other hand, he invited people to abandon their religious beliefs by denying the Creation, and thus undermined at all ethical values that might prove to be obstacles to the ruthlessness of the "struggle for survival."

Humanity has paid a heavy price in the 20th century for the dissemination of these callous views which lead people to acts of ruthlessness and cruelty.

What 'The Law of the Jungle' Led to: Fascism

As Darwinism fed racism in the 19th century, it formed the basis of an ideology that would develop and drown the world in blood in the 20th century: Nazism.

A strong Darwinist influence can be seen in Nazi ideologues. When one examines this theory, which was given shape by Adolf Hitler and Alfred Rosenberg, one comes across such concepts as "natural selection", "selective mating", and "the struggle for survival between the races", which are repeated dozens of time in the works of Darwin. When calling his book *Mein Kampf* (My Struggle), Hitler was inspired by the Darwinist struggle for survival and the principle that victory went to the fittest. He particularly talks about the struggle between the races:

> History would culminate in a new millennial empire of unparalleled splendour, based on a new racial hierarchy ordained by nature herself.4

In the 1933 Nuremberg party rally, Hitler proclaimed that "a higher race subjects to itself a lower race... a right which we see in nature and which can be regarded as the sole conceivable right".

That the Nazis were influenced by Darwinism is a fact that almost all historians who are expert in the matter accept. The historian Hickman describes Darwinism's influence on Hitler as follows:

> (Hitler) was a firm believer and preacher of evolution. Whatever the deeper, profound, complexities of his psychosis, it is certain that [the concept of struggle was important because]... his book, Mein Kampf, clearly set forth a number of evolutionary ideas, particularly those emphasizing struggle, survival of the fittest and the extermination of the weak to produce a better society.5

Hitler, who emerged with these views, dragged the world to violence that had never before been seen. Many ethnic and political groups, and especially the Jews, were exposed to terrible cruelty and slaughter in the Nazi concentration camps. World War II, which began with the Nazi invasion, cost 55 million lives. What lay behind the greatest tragedy in world history was Darwinism's concept of the "struggle for survival."

The Bloody Alliance: Darwinism and Communism

While fascists are found on the right wing of Social Darwinism, the left wing is occupied by communists. Communists have always been among the fiercest defenders of Darwin's theory.

This relationship between Darwinism and communism goes right back to the founders of both these "isms". Marx and Engels, the founders of communism, read Darwin's *The Origin of Species* as soon as it came out, and were amazed at its 'dialectical materialist' attitude. The correspondence between Marx and Engels showed that they saw Darwin's theory as "containing the basis in natural history for communism". In his book *The Dialectics of Nature*, which he wrote under the influence of Darwin, Engels was full of praise for Darwin, and tried to make his own contribution to the theory in the chapter "The Part Played by Labour in the Transition from Ape to Man".

Russian communists who followed in the footsteps of Marx and Engels, such as Plekhanov, Lenin, Trotsky and Stalin, all agreed with Darwin's theory of evolution. Plekhanov, who is seen as the founder of Russian communism, regarded **marxism as "Darwinism in its application to social science"**.[6]

Trotsky said, **"Darwin's discovery is the highest triumph of the dialectic in the whole field of organic matter."** [7]

'Darwinist education' had a major role in the formation of communist cadres. For instance, historians note the fact that **Stalin was religious in his youth, but became an atheist primarily because of Darwin's books.**[8]

Mao, who established communist rule in China and killed millions of people, openly stated that **"Chinese socialism is founded upon Darwin and the theory of evolution."** [9]

The Harvard University historian James Reeve Pusey goes into great detail regarding Darwinism's effect on Mao and Chinese communism in his research book *China and Charles Darwin*.[10]

In short, there is an unbreakable link between the theory of evolution and communism. The theory claims that living things are the product of chance, and provides a so-called scientific support for atheism. Communism, an atheist ideology, is for that reason firmly tied to Darwinism. Moreover, the theory of evolution proposes that development in nature is possible thanks to conflict (in other words "the struggle for survival") and supports the concept of "dialectics" which is fundamental to communism.

If we think of the communist concept of "dialectical conflict", which killed some 120 million people during the 20th century, as a "killing machine" then we can better understand the dimensions of the disaster that Darwinism visited on the planet.

Darwinism and Terrorism

As we have so far seen, Darwinism is at the root of various ideologies of violence that have spelled disaster to mankind in the 20th century. The fundamental concept behind this understanding and method is **"fighting whoever is not one of us."**

We can explain this in the following way: There are different beliefs, worldviews and philosophies in the world. It is very natural that all these diverse ideas have traits opposing one another. However, these different stances can look at each other in one of two ways:

1) They can respect the existence of those who are not like them and try to establish dialogue with them, employing a humane method. Indeed, this method conforms with the morality of the Qur'an.

2) They can choose to fight others, and to try to secure an advantage by damaging them, in other words, to behave like a wild animal. This is a method employed by materialism, that is, irreligion.

The horror we call terrorism is nothing other than a statement of the second view.

When we consider the difference between these two approaches, we can see that the idea of **"man as a fighting animal"** which Darwinism has subconsciously imposed on people is particularly influential. Individuals and groups who choose the way of conflict may never have heard of Darwinism and the principles of that ideology. But at the end of the day they agree with a view whose philosophical basis rests on Darwinism. What leads them to believe in the rightness of this view is such Darwinism-based slogans as "In this world, the strong survive", "Big fish swallow little ones", "War is a virtue", and "Man advances by waging war". Take Darwinism away, and these are nothing but empty slogans.

Actually, when Darwinism is taken away, no philosophy of 'conflict' remains. The three divine religions that most people in the world believe in, Islam, Christianity and Judaism, all oppose violence. All three religions wish to bring peace and harmony to the world, and oppose innocent people being killed and suffering cruelty and torture. Conflict and violence violate the morality that God has set out for man, and are abnormal and unwanted concepts. However, Darwinism sees and portrays conflict and violence as natural, justified and correct concepts that have to exist.

For this reason, if some people commit terrorism using the concepts and symbols of Islam, Christianity or Judaism in the name of those religions, you can be sure that those people are not Muslims, Christians or Jews. They are real Social Darwinists. They hide under a cloak of religion, but they are not genuine believers. Even if they claim to be serving religion, they are actually enemies of religion and of believers. That is because they are ruthlessly committing a crime that religion forbids, and in such a way as to blacken religion in peoples' eyes.

For this reason, the root of the terrorism that plagues our planet is not any of the divine religions, but in atheism, and the expression of atheism in our times: "Darwinism" and "materialism."

ISLAM IS NOT THE SOURCE OF TERRORISM, BUT ITS SOLUTION

Some people who say they are acting in the name of religion may misunderstand their religion or practice it wrongly. For that reason, it would be wrong to form ideas about that religion by taking these people as an example. The best way to understand a religion is to study its divine source.

The holy source of Islam is the Qur'an; and the model of morality in the Qur'an-Islam-is completely different from the image of it formed in the minds of some westerners. The Qur'an is based on the concepts of morality, love, compassion, mercy, humility, sacrifice, tolerance and peace, and a Muslim who lives by that morality in its true sense will be most polite, considerate, tolerant, trustworthy and accomodating. He will spread love, respect, harmony and the joy of living all around him.

Islam is a Religion of Peace and Well-Being

The word Islam is derived from the word meaning "peace" in Arabic. Islam is a religion revealed to mankind with the intention of presenting a peacable life through which the infinite compassion and mercy of God manifest on earth. God calls all people to Islamic morals through through which mercy, compassion, tolerance and peace can be experienced all over the world. In Surat al-Baqara verse 208, God addresses the believers as follows:

> You who believe! Enter absolutely into peace (Islam). Do not follow in the footsteps of Satan. He is an outright enemy to you.

As the verse makes clear, security can only be ensured by 'entering into Islam', that is, living by the values of the Qur'an.

God Has Condemned Wickedness

God has commanded people to avoid committing evil; He has forbidden disbelief, immorality, rebellion, cruelty, aggressiveness, murder and bloodshed. He describes those who fail to obey this command as "following in Satan's footsteps" and adopting a posture that is openly revealed to be sinful in the Qur'an. A few of the many verses on this matter in the Qur'an read:

> But as for those who break God's contract after it has been agreed and sever what God has commanded to be joined, and cause corruption in the earth, the curse will be upon them. They will have the Evil Abode. (Surat ar-Ra'd: 25)

> Seek the abode of the hereafter with what God has given you, without forgetting your portion of the world. And do good as God has been good to you. And do not seek to cause mischief on earth. God does not love mischief makers.' (Surat al-Qasas: 77)

As we can see, God has forbidden every kind of mischievous acts in the religion of Islam including terrorism and violence, and condemned those who commit such deeds. A Muslim lends beauty to the world and improves it.

Islam Defends Tolerance and Freedom of Speech

Islam is a religion which provides and guarantees freedom of ideas, thought and life. It has issued commands to prevent and forbid tension, disputes, slander and even negative thinking among people.

In the same way that it is determinedly opposed to terrorism and all acts of violence, it has also forbidden even the slightest ideological pressure to be put on them:

> There is no compulsion in religion. Right guidance has become clearly distinct from error. Anyone who rejects false gods and believes in God has grasped the Firmest Handhold, which will never give way. God is All-Hearing, All-Knowing. (Surat al-Baqara: 256)

> So remind, you need only to remind. You cannot compel them to believe. (Surat al-Ghashiyah: 22)

Forcing people to believe in a religion or to adopt its forms of belief is completely contrary to the essence and spirit of Islam. According to Islam, true faith is only possible with free will and freedom of conscience. Of course, Muslims can advise and encourage each other about the features of Qur'anic morality, but they will never resort to compulsion, nor any kind of physical or psychological pressure. Neither will they use any worldly privilege to turn someone towards religion.

Let us imagine a completely opposite model of society. For example, a world in which people are forced by law to practice religion. Such a model of society is completely contrary to Islam because **faith and worship are only of any value when they are directed to God by the free will of the individual**. If a system imposes belief and worship on people, then they will become religious only out of fear of that system. From the religious point of view, what really counts is that religion should be lived for God's good pleasure in an environment where peopls' consciences are totally free.

God Has Made the Killing of Innocent People Unlawful

According to the Qur'an, one of the greatest sins is to kill a human being who has committed no fault.

> **...If someone kills another person – unless it is in retaliation for someone else or for causing corruption in the earth – it is as if he had murdered all mankind. And if anyone gives life to another person, it is as if he had given life to all mankind. Our Messengers came to them with Clear Signs but even after that many of them committed outrages in the earth. (Surat al-Ma'ida: 32)**

> **Those who do not call on any other deity together with God and do not kill anyone God has made inviolate, except with the right to do so, and do not fornicate; anyone who does that will receive an evil punishment. (Surat al-Furqan: 68)**

As the verses suggest, a person who kills innocent people for no reason is threatened with a great torment. God has revealed that killing even a single person is as evil as murdering all mankind. A person who observes God's limits can do no harm to a single human, let alone massacre thousands of innocent people. Those who assume that they can avoid justice and thus punishment in this world will never succeed, for they will have to give an account of their deeds in the presence of God. That is why believers, who know that they will give an account of their deeds after death, are very meticulous to observe God's limits.

God Commands Believers to be Compassionate and Merciful

Islamic morality is described in one verse as:

> ...To be one of those who believe and urge each other to steadfastness and urge each other to compassion. Those are the Companions of the Right. (Surat al-Balad: 17-18)

As we have seen in this verse, one of the most important moral precepts that God has sent down to His servants so that they may receive salvation and mercy and attain Paradise, is to **"urge each other to compassion"**.

İslam as described in the Qur'an is a modern, enlightened, progressive religion. A Muslim is above all a person of peace; he is tolerant with a democratic spirit, cultured, enlightened, honest, knowledgeable about art and science and civilized.

A Muslim educated in the fine moral teaching of the Qur'an, approaches everyone with the love that İslam expects. He shows respect for every idea and he values art and aesthetics. He is conciliatory in the face of every event, diminishing tension and restoring amity. In societies composed of individuals such as this, there will be a more developed civilization, a higher social morality, more joy, happiness, justice, security, abundance and blessings than in the most modern nations of the world today.

God Has Commanded Tolerance and Forgiveness

The concept of forgiveness and tolerance, described in the words, **'Make allowences for people'** (Surat al-A'raf: 199), is one of the most fundamental tenets of İslam.

When we look at the history of İslam, the way that Muslims have translated this important feature of Qur'anic morality into the life of society can be seen quite clearly. Muslims have always brought with them an atmosphere of freedom and tolerance and destroyed unlawful practices wherever they have gone. They have enabled people whose religions, languages and cultures are completely different from one another to live together in peace and harmony under one roof, and provided peace and harmony for its own members. One of the most important reasons for the centuries-long existence of the Ottoman Empire, which spread over an enormous region, was the atmosphere of tolerance and understanding that

Islam brought with it. Muslims, who have been known for their tolerant and loving natures for centuries, have always been the most compassionate and just of people. Within this multi-national structure, all ethnic groups have been free to live according to their own religions, and their own rules.

True tolerance can only bring peace and well-being to the world when implemented along the lines set out in the Qur'an. Attention is drawn to this fact in a verse which reads:

> **A good action and a bad action are not the same. Repel the bad with something better and, if there is enmity between you and someone else, he will be like a bosom friend. (Surat al-Fussilat: 34)**

Conclusion

All of this shows that the morality that Islam recommends to mankind brings to the world the virtues of peace, harmony and justice. The barbarism known as terrorism, that is so preoccupying the world at present, is the work of ignorant and fanatical people, completely estranged from Qur'anic morality, and who have absolutely nothing to do with religion. The solution to these people and groups who try to carry out their savagery under the mask of religion is the teaching of true Qur'anic morality. In other words, Islam and Qur'anic morality are solutions to the scourge of terrorism, not supporters of it.

---

1. Charles Darwin, *The Descent of Man*, 2nd edition, New York, A L. Burt Co., 1874, p. 178
2. Lalita Prasad Vidyarthi, *Racism, Science and Pseudo-Science*, Unesco, France, Vendôme, 1983. p. 54
3. Theodore D. Hall, *The Scientific Background of the Nazi "Race Purification" Program*, http://www.trufax.org/avoid/nazi.html
4. L.H. Gann, "Adolf Hitler, The Complete Totalitarian", *The Intercollegiate Review*, Fall 1985, p. 24; cited in Henry M. Morris, *The Long war Against God*, Baker Book House, 1989, p. 78
5. Hickman, R., *Biocreation*, Science Press, Worthington, OH, pp. 51–52, 1983; Jerry Bergman, "Darwinism and the Nazi Race Holocaust", *Creation Ex Nihilo Technical Journal* 13 (2): 101-111, 1999
6. Robert M. Young, *Darwinian Evolution and Human History*, Historical Studies on Science and Belief, 1980
7. Alan Woods and Ted Grant, *Reason in Revolt: Marxism and Modern Science*, London: 1993
8. Alex de Jonge, *Stalin and The Shaping of the Soviet Uninon*, William Collins Sons & Limited Co., Glasgow, 1987, p. 22
9. K. Mehnert, *Kampf um Mao's Erbe*, Deutsche Verlags-Anstalt, 1977
10. James Reeve Pusey, *China and Charles Darwin*, Cambridge, Massachusetts, 1983

PART I

THE REFUTATION OF DARWINISM

INTRODUCTION

Why the Theory of Evolution?

For some people the theory of evolution or Darwinism has only scientific connotations, with seemingly no direct implication in their daily lives. This is, of course, a common misunderstanding. Far beyond just being an issue within the framework of the biological sciences, the theory of evolution constitutes the underpinning of a deceptive philosophy that has held sway over a large number of people: Materialism.

Materialist philosophy, which accepts only the existence of matter and presupposes man to be 'a heap of matter', asserts that he is no more than an animal, with 'conflict' the sole rule of his existence. Although propagated as a modern philosophy based on science, materialism is in fact an ancient dogma with no scientific basis. Conceived in Ancient Greece, the dogma was rediscovered by the atheistic philosophers of the 18th century. It was then implanted in the 19th century into several science disciplines by thinkers such as Karl Marx, Charles Darwin and Sigmund Freud. In other words science was distorted to make room for materialism.

The past two centuries have been a bloody arena of materialism: Ideologies based on materialism (or competing ideologies arguing against materialism, yet sharing its basic tenets) have brought permanent violence, war and chaos to the world. Communism, responsible for the death of 120 million people, is the direct outcome of materialistic philosophy. Fascism, despite pretending to be an alternative to the materialistic world-view, accepted the fundamental materialist concept of progress though conflict and sparked off oppressive regimes, massacres, world wars and genocide.

Besides these two bloody ideologies, individual and social ethics have also been corrupted by materialism.

The deceptive message of materialism, reducing man to an animal whose existence is coincidental and with no responsibility to any being, demolished moral pillars such as love, mercy, self-sacrifice, modesty, honesty and justice. Having been misled by the materialists' motto "life is a struggle", people came to see their lives as nothing more than a clash of in-

Karl Marx made it clear that Darwin's theory provided a solid ground for materialism and thus also for communism. He also showed his sympathy to Darwin by dedicating *Das Kapital*, which is considered as his greatest work, to him. In the German edition of the book, he wrote: "From a devoted admirer to Charles Darwin"

terests which, in turn, led to life according to the law of the jungle.

Traces of this philosophy, which has a lot to answer as regards man-made disasters of the last two centuries, can be found in every ideology that perceives differences among people as a 'reason for conflict'. That includes the terrorists of the present day who claim to uphold religion, yet commit one of the greatest sins by murdering innocent people.

The theory of evolution, or Darwinism, comes in handy at this point by completing the jigsaw puzzle. It provides the myth that materialism is a scientific idea. That is why, Karl Marx, the founder of communism and dialectical materialism, wrote that Darwinism was "the basis in natural history" for his worldview.[1]

However, that basis is rotten. Modern scientific discoveries reveal over and over again that the popular belief associating Darwinism with science is false. Scientific evidence refutes Darwinism comprehensively and reveals that the origin of our existence is not evolution but creation. God has created the universe, all living things and man.

This book has been written to make this fact known to people. Since its first publication, originally in Turkey and then in many other countries, millions of people have read and appreciated the book. In addition to Turkish, it has been printed in English, Italian, Spanish, Russian, Bosnian, Arabic, Malay and Indonesian. (The text of the book is freely available in all these languages at www.evolutiondeceit.com.)

The impact of *The Evolution Deceit* has been acknowledged by standard-bearers of the opposing view. Harun Yahya was the subject of a *New Scientist* article called "Burning Darwin". This leading popular Darwinist periodical noted in its 22 April 2000 issue that Harun Yahya "is an interna-

tional hero" sharing its concern that his books "have spread everywhere in the Islamic world."

Science, the leading periodical of the general scientific community, emphasized the impact and sophistication of Harun Yahya's works. The *Science* article "Creationism Takes Root Where Europe, Asia Meet", dated 18 May 2001, observed that in Turkey "sophisticated books such as *The Evolution Deceit* and *The Dark Face of Darwinism...* have become more influential than textbooks in certain parts of the country". The reporter then goes on to assess Harun Yahya's work, which has initiated "one of the world's strongest anti-evolution movements outside of North America".

Although such evolutionist periodicals note the impact of *The Evolution Deceit*, they do not offer any scientific replies to its arguments. The reason is, of course, that it is simply not possible. The theory of evolution is in complete deadlock, a fact you will discover as you read the following chapters. The book will help you realise that Darwinism is not a scientific theory but a pseudo-scientific dogma upheld in the name of materialist philosophy, despite counter evidence and outright refutation.

It is our hope that *The Evolution Deceit* will for a long time continue its contribution towards the refutation of materialist-Darwinist dogma which has been misleading humanity since the 19th century. And it will remind people of the crucial facts of our lives, such as how we came into being and what our duties to our Creator are.

FOREWORD

The Greatest Miracle of Our Times: Belief in the Evolution Deceit

All the millions of living species on the earth possess miraculous features, unique behavioural patterns and flawless physical structures. Every one of these living things has been created with its own unique detail and beauty. Plants, animals, and man above all, were all created with great knowledge and art, from their external appearances down to their cells, invisible to the naked eye. Today there are a great many branches of science, and tens of thousands of scientists working in those branches, that research every detail of those living things, uncover the miraculous aspects of those details and try to provide an answer to the question of how they came into being.

Some of these scientists are astonished as they discover the miraculous aspects of these structures they study and the intelligence behind that coming into existence, and they witness the infinite knowledge and wisdom involved. Others, however, surprisingly claim that all these miraculous features are the product of blind chance. These scientists believe in the theory of evolution. In their view, the proteins, cells and organs that make up these living things all came about by a string of coincidences. It is quite amazing that such people, who have studied for long years, carried out lengthy studies and written books about the miraculous functioning of just one organelle within the cell, itself too small to be seen with the naked eye, can think that these extraordinary structures came about by chance.

The chain of coincidences such eminent professors believe in so flies in the face of reason that their doing so leaves outside observers utterly amazed. According to these professors, a number of simple chemical substances first came together and formed a protein - which is no more possible than a randomly scattered collection of letters coming together to form a poem. Then, other coincidences led to the emergence of other proteins. These then also combined by chance in an organised manner. Not just proteins, but DNA, RNA, enzymes, hormones and cell organelles, all of which

The Greatest Miracle of Our Times: Belief in the Evolution Deceit

are very complex structures within the cell, coincidentally happened to emerge and come together. As a result of these billions of coincidences, the first cell came into being. The miraculous ability of blind chance did not stop there, as these cells then just happened to begin to multiply. According to the claim in question, another coincidence then organised these cells and produced the first living thing from them.

Billions of "chance events" had to take place together for just one eye in a living thing to form. Here too the blind process known as coincidence entered the equation: It first opened two holes of the requisite size and in the best possible place in the skull, and then cells that happened by chance to find themselves in those places coincidentally began to construct the eye.

As we have seen, coincidences acted in the knowledge of what they wanted to produce. Right from the very start, "chance" knew what seeing, hearing and breathing were, even though there was not one example of such things anywhere in the world at that time. It displayed great intelligence and awareness, exhibited considerable forward planning, and constructed life step by step. This is the totally irrational scenario to which these professors, scientists and researchers whose names are greatly respected and whose ideas are so influential have devoted themselves. Even now, with a childish stubbornness, they exclude anyone who refuses to believe in such fairy tales, accusing them of being unscientific and bigoted. There is really no difference between this and the bigoted, fanatical and ignorant medieval mentality that punished those who claimed the earth was not flat.

What is more, some of these people claim to be Muslims and believe in God. Such people find saying, "God created all of life" unscientific, and yet are quite able to believe instead that saying, "It came about in an unconscious process consisting of billions of miraculous coincidences" is scientific.

If you put a carved stone or wooden idol in front of these people and told them, "Look, this idol created this room and everything in it" they would say that was utterly stupid and refuse to believe it. Yet despite that

The mentality of those who claim that life formed from nonliving matter by random gradual changes and who defend this with a childish stubbornness despite all scientific evidence to the contrary, is no different from the bigoted, fanatical and ignorant medieval mentality that punished those who claimed the earth was not flat.

they declare the nonsense that "The unconscious process known as chance gradually brought this world and all the billions of wonderful living things in it into being with enormous planning" to be the greatest scientific explanation.

In short, these people regard chance as a god, and claim that it is intelligent, conscious and powerful enough to create living things and all the sensitive balances in the universe. When told that it was God, the possessor of infinite wisdom, who created all living things, these evolutionist professors refuse to accept the fact, and maintain that unconscious, unintelligent, powerless billions of coincidences with no will of their own are actually a creative force.

The fact that educated, intelligent and knowledgeable people can as a group believe in the most irrational and illogical claim in history, as if under a spell, is really a great miracle. In the same way that God miraculously creates something like the cell, with its extraordinary organization and properties, this people are just as miraculously so blind and devoid of understanding as to be unable to see what is under their very noses. It is one of God's miracles that evolutionists are unable to see facts that even tiny children can, and fail to grasp them no matter how many times they are told.

You will frequently come across that miracle as you read this book. And you will also see that as well as being a theory that has totally collapsed in the face of the scientific facts, Darwinism is a great deceit that is utterly incompatible with reason and logic, and which belittles those who defend it.

CHAPTER 1

To Be Freed From Prejudice

Most people accept everything they hear from scientists as strictly true. It does not even occur to them that scientists may also have various philosophical or ideological prejudices. The fact of the matter is that evolutionist scientists impose their own prejudices and philosophical views on the public under the guise of science. For instance, although they are aware that random events do not cause anything other than irregularity and confusion, they still claim that the marvellous order, plan, and design seen both in the universe and in living organisms arose by chance.

For instance, such a biologist easily grasps that there is an incomprehensible harmony in a protein molecule, the building block of life, and that there is no probability that this might have come about by chance. Nevertheless, he alleges that this protein came into existence under primitive earth conditions by chance billions of years ago. He does not stop there; he also claims, without hesitation, that not only one, but millions of proteins formed by chance and then incredibly came together to create the first living cell. Moreover, he defends his view with a blind stubbornness. This person is an "evolutionist" scientist.

If the same scientist were to find three bricks resting on top of one another while walking along a flat road, he would never suppose that these bricks had come together by chance and then climbed up on top of each other, again by chance. Indeed, anyone who did make such an assertion would be considered insane.

How then can it be possible that people who are able to assess ordinary events rationally can adopt such an irrational attitude when it comes to thinking about their own existence?

It is not possible to claim that this attitude is adopted in the name of science: scientific approach requires taking both alternatives into consideration wherever there are two alternatives equally possible concerning a certain case. And if the likelihood of one of the two alternatives is much lower, for example if it is only one percent, then the rational and scientific thing to do is to consider the other alternative, whose likelihood is 99 percent, to be the valid one.

Let us continue, keeping this scientific basis in mind. There are two views that can be set forth regarding how living beings came into being on earth. The first is that God creates all living beings in their present complex structure. The second is that life was formed by unconscious, random coincidences. The latter is the claim of the theory of evolution.

When we look at the scientific data, that of molecular biology for instance, we can see that there is no chance whatsoever that a single living cell-or even one of the millions of proteins present in this cell-could have come into existence by chance as the evolutionists claim. As we will illustrate in the following chapters, probabilistic calculations also confirm this many times over. So the evolutionist view on the emergence of living beings has zero probability of being true.

This means that the first view has a "one hundred percent" probability of being true. That is, life has been consciously brought into being. To put it in another way, it was "created". All living beings have come into existence by the design of a Creator exalted in superior power, wisdom, and knowledge. This reality is not simply a matter of conviction; it is the normal conclusion that wisdom, logic and science take one to.

Under these circumstances, our "evolutionist" scientist ought to withdraw his claim and adhere to a fact that is both obvious and proven. To do otherwise is to demonstrate that he is actually someone who is sacrificing science on behalf of his philosophy, ideology, and dogma rather than being a true scientist.

The anger, stubbornness, and prejudices of our "scientist" increase more and more every time he confronts reality. His attitude can be explained with a single word: "faith". Yet it is a blind superstitious faith, since there can be no other explanation for one's disregard of all the facts or for a lifelong devotion to the preposterous scenario that he has constructed in his imagination.

Blind Materialism

The faith that we are talking about is the **materialistic philosophy**, which argues that matter has existed for all eternity and there is nothing other than matter. The theory of evolution is the so-called "scientific foundation" for this materialistic philosophy and that theory is blindly defended in order to uphold this philosophy. When science invalidates the claims of evolution-and that is the very point that has been reached here at the end of the 20th century-it then is sought to be distorted and brought into a position where it supports evolution for the sake of keeping materialism alive.

A few lines written by one of the prominent evolutionist biologists of Turkey is a good example that enables us to see the disordered judgement and discretion that this blind devotion leads to. This scientist discusses the probability of the coincidental formation of Cytochrome-C, which is one of the most essential enzymes for life, as follows:

> The probability of the formation of a Cytochrome-C sequence is as likely as zero. That is, if life requires a certain sequence, it can be said that this has a probability likely to be realised once in the whole universe. Otherwise, some metaphysical powers beyond our definition should have acted in its formation. To accept the latter is not appropriate to the goals of science. We therefore have to look into the first hypothesis.[2]

This scientist finds it "more scientific" to accept a possibility "as likely as zero" rather than creation. However according to the rules of science, if there are two alternative explanations concerning an event and if one of them has "as likely as zero" a possibility of realisation, then the other one is the correct alternative. However the **dogmatic materialistic approach forbids the admittance of a superior Creator**. This prohibition drives this scientist-and many others who believe in the same materialist dogma-to accept claims that are completely contrary to reason.

People who believe and trust these scientists also become enthralled and blinded by the same materialistic spell and they adopt the same insensible psychology when reading their books and articles.

This dogmatic materialistic point of view is the reason why many prominent names in the scientific community are atheists. Those who free themselves from the thrall of this spell and think with an open mind do not hesitate to accept the existence of a Creator. American biochemist Dr Michael J. Behe, one of those prominent names who support the theory of "**intelligent design**" that has lately become very accepted, describes the scientists who resist believing in the "design" or "creation" of living organisms thus:

> Over the past four decades, modern biochemistry has uncovered the secrets of the cell. It has required tens of thousands of people to dedicate the better parts of their lives to the tedious work of the laboratory... The result of these cumulative efforts to investigate the cell- to investigate life at the molecular level-is a loud, clear, piercing cry of "design!". The result is so unambiguous and so significant that it must be ranked as one of the greatest achievements in the history of science... Instead a curious, embarrassed silence surrounds the stark complexity of the cell. Why does the scientific community not greedily embrace its startling discovery? Why is the observation of design handled with intellectual gloves? The dilemma is that while one side of the [issue] is labelled intelligent design, the other side must be labelled God.3

Michael Behe:
"An embarrased silence surrounds the stark complexity of the cell"

This is the predicament of the atheist evolutionist scientists you see in magazines and on television and whose books you may be reading. All the scientific research carried out by these people demonstrates to them the existence of a Creator. Yet they have become so insensitised and blinded by the dogmatic materialist education they have absorbed that they still persist in their denial.

People who steadily neglect the clear signs and evidences of the Creator become totally insensitive. Caught up in an ignorant self-confidence caused by their insensitivity, they may even end up supporting an absurdity as a virtue. A good case in point is the prominent evolutionist Richard Dawkins who calls upon Christians not to assume that they have witnessed a miracle even if they see the statue of the Virgin Mary wave to them. According to Dawkins, "Perhaps all the atoms of the statue's arm just happened to move in the same direction at once-a low probability event to be sure, but possible." 4

The psychology of the unbeliever has existed throughout history. In the Qur'an it is described thus:

Even if We did send unto them angels, and the dead did speak unto them, and We gathered together all things before their very eyes, they are not the ones to believe, unless it is in God's plan. But most of them ignore (the truth). (Surat al-Anaam : 111)

As this verse makes clear, the dogmatic thinking of the evolutionists is not an original way of thinking, nor is it even peculiar to them. In fact, what the evolutionist scientist maintains is not a modern scientific thought but an ignorance that has persevered since the most uncivilised pagan communities.

Richard Dawkins, busy with propagating evolution

The same psychology is defined in another verse of the Qur'an:

Even if We opened out to them a gate from heaven and they were to continue (all day) ascending therein, they would only say: "Our eyes have been intoxicated: Nay, we have been bewitched by sorcery." (Surat Al-Hijr : 14-15)

Mass Evolutionist Indoctrination

As indicated in the verses cited above, one of the reasons why people cannot see the realities of their existence is a kind of "spell" impeding their reasoning. It is the same "spell" that underlies the world-wide acceptance of the theory of evolution. What we mean by spell is a conditioning acquired by indoctrination. People are exposed to such an intense indoctrination about the correctness of the theory of evolution that they often do not even realise the distortion that exists.

This indoctrination creates a negative effect on the brain and disables the faculty of judgement. Eventually, the brain, being under a continuous indoctrination, starts to perceive the realities not as they are but as they have been indoctrinated. This phenomenon can be observed in other examples. For instance, if someone is hypnotised and indoctrinated that the bed he is lying on is a car, he perceives the bed as a car after the hypnosis session. He thinks that this is very logical and rational because he really sees it that way and has no doubt that he is right. Such examples as the one above, which show the efficiency and the power of the mechanism of indoctrination, are scientific realities that have been verified by countless experiments that have been reported in the scientific literature and are the everyday fare of psychology and psychiatry textbooks.

The theory of evolution and the materialistic world view that relies on it are imposed on the masses by such indoctrination methods. People who continuously encounter the indoctrination of evolution in the media, academic sources, and "scientific" platforms, fail to realise that accepting this theory is in fact contrary to the most basic principles of reason. The same indoctrination captures scientists as well. Young names stepping up in their scientific careers adopt the materialist world view more and more as time passes. Enchanted by this spell, many evolutionist scientists go on searching for scientific confirmation of 19th century's irrational and outdated evolutionist claims that have long since been refuted by scientific evidence.

There are also additional mechanisms that force scientists to be evolutionist and materialist. In Western countries, a scientist has to observe some standards in order to be promoted, to receive academic recognition, or to have his articles published in scientific journals. A straightforward acceptance of evolution is the number-one criterion. This system drives these scientists so far as to spend their whole lives and scientific careers for the sake of a dogmatic belief. American molecular biologist Jonathan Wells refers to these pressure mechanisms in his book *Icons of Evolution* published in 2000:

> ...Dogmatic Darwinists begin by imposing a narrow interpretation on the evidence and declaring it the only way to do science. Critics are then labeled unscientific; their articles are rejected by mainstream journals, whose editorial boards are dominated by the dogmatists; the critics are denied funding by government agencies, who send grant proposals to the dogmatists for "peer" review; and eventually the critics are hounded out of scientific community altogether. In the process, evidence against the Darwinian view simply disappears, like witnesses against the Mob. Or the evidence is buried in specialized publications, where only a dedicated researcher can find. Once critics have been silenced and counter-evidence has been buried, the dogmatists announce that there is scientific debate about their theory, and no evidence against it.[5]

This is the reality that continues to lie behind the assertion "evolution is still accepted by the world of science". Evolution is kept alive not because it has a scientific worth but because it is an ideological obligation. Very few of the scientists who are aware of this fact can risk pointing out that the king isn't wearing any clothes.

In the rest of this book, we will be reviewing the findings of modern science against evolution which are either disregarded by evolutionists or "buried in specialized publications", and display of the clear evidence of God's existence. The reader will witness that evolution theory is in fact a deceit-a deceit that is belied by science at every step but is upheld to veil the fact of creation. What is to be hoped of the reader is that he will wake up from the spell that blinds people's minds and disrupts their ability to judge and he will reflect seriously on what is related in this book.

If he rids himself of this spell and thinks clearly, freely, and without any prejudice, he will soon discover the crystal-clear truth. This inevitable truth, also demonstrated by modern science in all its aspects, is that living organisms came into existence not by chance but as a result of creation. Man can easily see the fact of creation when he considers how he himself exists, how he has come into being from a drop of water, or the perfection of every other living thing.

CHAPTER 2

A Brief History of the Theory

The roots of evolutionist thought go back as far as antiquity as a dogmatic belief attempting to deny the fact of creation. Most of the pagan philosophers in ancient Greece defended the idea of evolution. When we take a look at the history of philosophy we see that the idea of evolution constitutes the backbone of many pagan philosophies.

However, it is not this ancient pagan philosophy, but faith in God which has played a stimulating role in the birth and development of modern science. Most of the people who pioneered modern science believed in the existence of God; and while studying science, they sought to discover the universe God has created and to perceive His laws and the details in His creation. Astronomers such as **Copernicus, Keppler,** and **Galileo**; the father of paleontology, **Cuvier**; the pioneer of botany and zoology, **Linnaeus**; and **Isaac Newton**, who is referred to as the "greatest scientist who ever lived", all studied science believing not only in the existence of God but also that the whole universe came into being as a result of His creation.[6] **Albert Einstein**, considered to be the greatest genius of our age, was another devout scientist who believed in God and stated thus; "I cannot conceive of a genuine scientist without that profound faith. The situation may be expressed by an image: science without religion is lame."[7]

One of the founders of modern physics, German physician **Max Planck** said: "Anybody who has been seriously engaged in scientific work of any kind realizes that over the entrance to the gates of the temple of science are written the words: Ye must have faith. It is a quality which the scientist cannot dispense with."[8]

The theory of evolution is the outcome of the materialist philosophy that surfaced with the reawakening of ancient materialistic philosophies and became widespread in the 19th century. As we have indicated before, materialism seeks to explain nature through purely material factors. Since it denies creation right from the start, it asserts that every thing, whether animate or inanimate, has appeared without an act of creation but rather as a result of a coincidence that then acquired a condition of order. The human mind however is so structured as to comprehend the existence of an organising will wherever it sees order. Materialistic philosophy, which is contrary to this very basic characteristic of the human mind, produced "the theory of evolution" in the middle of the 19th century.

Darwin's Imagination

The person who put forward the theory of evolution the way it is defended today, was an amateur English naturalist, Charles Robert Darwin.

Darwin had never undergone a formal education in biology. He took only an amateur interest in the subject of nature and living things. His interest spurred him to voluntarily join an expedition on board a ship named H.M.S. Beagle that set out from England in 1832 and travelled around different regions of the world for five years. Young Darwin was greatly impressed by various living species, especially by certain finches that he saw in the Galapagos Islands. He thought that the variations in their beaks were caused by their adaptation to their habitat. With this idea in mind, he supposed that the origin of life and species lay in the concept of "adaptation to the environment". According to Darwin, different living species were not created separately by God but rather came from a common ancestor and became differentiated from each other as a result of natural conditions.

Darwin's hypothesis was not based on any scientific discovery or experiment; in time however he turned it into a pretentious theory with the support and encouragement he received from the famous materialist biologists of his time. The idea was that the individuals that adapted to the habitat in the best way transferred their qualities to subsequent generations; these advantageous qualities accumulated in time and transformed the individual into a species totally different from its ancestors. (The origin of these "advantageous qualities" was unknown at the time.) According to Darwin, man was the most developed outcome of this mechanism.

Darwin called this process **"evolution by natural selection"**. He thought he had found the "origin of species": the origin of one species was another species. He published these views in his book titled *The Origin of Species, By Means of Natural Selection* in 1859.

Darwin was well aware that his theory faced lots of problems. He confessed these in his book in the chapter "**Difficulties of the Theory**". These difficulties primarily consisted of the fossil record, complex organs of living things that could not possibly be explained by coincidence (e.g. the eye), and the instincts of living beings. Darwin hoped that these difficulties would be overcome by new discoveries; yet this did not stop him from coming up with a number of very inadequate explanations for some. The American physicist Lipson made the following comment on the "difficulties" of Darwin:

> On reading *The Origin of Species*, I found that Darwin was much less sure himself than he is often represented to be; the chapter entitled "Difficulties of the Theory" for example, shows considerable self-doubt. As a physicist, I was particularly intrigued by his comments on how the eye would have arisen.[9]

While developing his theory, Darwin was impressed by many evolutionist biologists preceding him, and primarily by the French biologist, **Lamarck**.[10] According to Lamarck, living creatures passed the traits they acquired during their lifetime from one generation to the next and thus evolved. For instance, giraffes evolved from antelope-like animals by extending their necks further and further from generation to generation as they tried to reach higher and higher branches for food. Darwin thus employed the thesis of "passing the acquired traits" proposed by Lamarck as the factor that made living beings evolve.

But both Darwin and Lamarck were mistaken because in their day, life could only be studied with very primitive technology and at a very inadequate level. Scientific fields such as genetics and biochemistry did not exist even in name. Their theories therefore had to depend entirely on their powers of imagination.

While the echoes of Darwin's book reverberated, an Austrian botanist by the name of **Gregor Mendel** discovered the laws of inheritance in 1865. Not much heard of until the end of the century, Mendel's discovery gained great importance in the early 1900s. This was the birth of the science of **genetics**. Somewhat later, the structure of the genes and the chromosomes was discovered. The discovery, in the 1950s, of the structure of the DNA molecule that incorporates genetic information threw the theory of evolution into a great crisis. The rea-

Charles Darwin

Darwin's Racism

One of the most important yet least-known aspects of Darwin is his racism: Darwin regarded white Europeans as more "advanced" than other human races. While Darwin presumed that man evolved from ape-like creatures, he surmised that some races developed more than others and that the latter still bore simian features. In his book, *The Descent of Man*, which he published after The Origin of Species, he boldly commented on "the greater differences between men of distinct races".[1] In his book, Darwin held blacks and Australian Aborigines to be equal to gorillas and then inferred that these would be "done away with" by the "civilised races" in time. He said:

> At some future period, not very distant as measured by centuries, the civilized races of man will almost certainly exterminate and replace the savage races throughout the world. At the same time the anthropomorphous apes... will no doubt be exterminated. The break between man and his nearest allies will then be wider, for it will intervene in a more civilised state, as we may hope, even than the Caucasian, and some ape as low as baboon, instead of as now between the negro or Australian and the gorilla.[2]

Darwin's nonsensical ideas were not only theorised, but also brought into a position where they provided the most important "scientific ground" for racism. Supposing that living beings evolved in the struggle for life, Darwinism was even adapted to the social sciences, and turned into a conception that came to be called "Social Darwinism".

Social Darwinism contends that existing human races are located at different rungs of the "evolutionary ladder", that the European races were the most "advanced" of all, and that many other races still bear "simian" features.

1- Benjamin Farrington, *What Darwin Really Said*. London: Sphere Books, 1971, pp. 54-56
2- Charles Darwin, *The Descent of Man*, 2nd ed., New York: A.L. Burt Co., 1874, p. 178

son was the incredible complexity of life and the invalidity of the evolutionary mechanisms proposed by Darwin.

These developments ought to have resulted in Darwin's theory being banished to the dustbin of history. However, it was not, because certain circles insisted on revising, renewing, and elevating the theory to a scientific platform. These efforts gain meaning only if we realise that behind the theory lay ideological intentions rather than scientific concerns.

The Desperate Efforts of Neo-Darwinism

Darwin's theory entered into a deep crisis because of the laws of genetics discovered in the first quarter of the 20th century. Nevertheless, a group of scientists who were determined to remain loyal to Darwin endeavoured to come up with solutions. They came together in a meeting organised by the Geological Society of America in 1941. Geneticists such as G. Ledyard Stebbins and Theodosius Dobzhansky, zoologists such as Ernst Mayr and Julian Huxley, paleontologists such as George Gaylord Simpson and Glenn L. Jepsen, and mathematical geneticists such as Ronald Fisher and Sewall Right, after long discussions, finally agreed on ways to "patch up" Darwinism.

This cadre focused on the question of the **origin of the advantageous variations that supposedly caused living organisms to evolve**-an issue that Darwin himself was unable to explain but simply tried to side-step by depending on Lamarck. The idea was now "**random mutations**". They named this new theory "**The Modern Synthetic Evolution Theory**", which was formulated by adding the concept of mutation to Darwin's natural selection thesis. In a short time, this theory came to be known as "**neo-Darwinism**" and those who put forward the theory were called "neo-Darwinists".

The following decades were to become an era of desperate attempts to prove neo-Darwinism. It was already known that **mutations**-or "accidents" -that took place in the genes of living organisms were always harmful. Neo-Darwinists tried to establish a case for "advantageous mutation" by carrying out thousands of mutation experiments. All their attempts ended in complete failure.

They also tried to prove that the first living organisms could have originated by chance under primitive terrestrial conditions that the theory posited but the same failure attended these experiments too. Every experiment that sought to prove that life could be generated by chance failed. Probability calculations prove that not even a single protein, the building-blocks of life, could have originated by chance. And the cell-which supposedly emerged by chance under primitive and uncontrolled terrestrial conditions according to the evolutionists-could not be synthesised by even the most sophisticated laboratories of the 20th century.

Neo-Darwinist theory is also defeated by **the fossil record**. No "transitional forms", which were supposed to show the gradual evolution of living organisms from primitive to advanced species as the neo-Darwinist theory claimed, have ever been found anywhere in the world. At the same time, comparative anatomy revealed that species that were supposed to have evolved from one another had in fact very different anatomical features and that they could never have been ancestors or descendants of each other.

But neo-Darwinism was never a scientific theory anyway, but was an ideological dogma if not to say some sort of "religion". The Darwinist professor of philosophy and zoology Michael Ruse confesses this in these words:

> And certainly, there's no doubt about it, that in the past, and I think also in the present, for many evolutionists, evolution has functioned as something with elements which are, let us say, akin to being a secular religion ... And it seems to me very clear that at some very basic level, evolution as a scientific theory makes a commitment to a kind of naturalism...[11]

This is why the champions of the theory of evolution still go on defending it in spite of all the evidence to the contrary. One thing they cannot agree on however is which of the different models proposed for the realisation of evolution is the "right" one. One of the most important of these models is the fantastic scenario known as "punctuated equilibrium".

Trial and Error: Punctuated Equilibrium

Most of the scientists who believe in evolution accept the neo-Darwinist theory of slow, gradual evolution. In recent decades, however, a different model has been proposed. Called "punctuated equilibrium", this model maintains that living species came about not through a series of small changes, as Darwin had maintained, but by sudden, large ones.

The Primitive Level of Science in Darwin's Time

FOCUS

When Darwin put forward his assumptions, the disciplines of genetics, microbiology, and biochemistry did not yet exist. If they had been discovered before Darwin put forward his theory, Darwin might easily have recognised that his theory was totally unscientific and might not have attempted to advance such meaningless claims. The information determining the species already exists in the genes and it is impossible for natural selection to produce new species through alterations in the genes.

Similarly, the world of science in those days had a very shallow and crude understanding of the structure and functions of the cell. If Darwin had had the chance to view the cell with an electron microscope, he would have witnessed the great complexity and extraordinary structure in the organelles of the cell. He would have beheld with his own eyes that it would not be possible for such an intricate and complex system to occur through minor variations. If he had known about bio-mathematics, then he would have realised that not even a single protein molecule, let alone a whole cell, could not have come into existence by chance.

Detailed studies of the cell were only possible after the discovery of the electron microscope. In Darwin's time, with the primitive microscopes seen here, it was only possible to view the outside surface of the cell

The first vociferous defenders of this notion appeared at the beginning of the 1970s. Two American paleontologists, **Niles Eldredge** and **Stephen Jay Gould**, were well aware that the claims of the neo-Darwinist theory were absolutely refuted by the fossil record. Fossils proved that living organisms did not originate by gradual evolution, but appeared suddenly and fully-formed. Neo-Darwinists were living with the fond hope-they still do-that the lost transitional forms

Stephen Jay Gould

would one day be found. Realising that this hope was groundless, Eldredge and Gould were nevertheless unable to abandon their evolutionary dogma, so they put forward a new model: punctuated equilibrium. This is the claim that evolution did not take place as a result of minor variations but rather in sudden and great changes.

This model was nothing but a model for fantasies. For instance, European paleontologist O.H. Shindewolf, who led the way for Eldredge and Gould, claimed that the first bird came out of a reptile egg, as a "gross mutation", that is, as a result of a huge "accident" that took place in the genetic structure.[12] According to the same theory, some land-dwelling animals could have turned into giant whales having undergone a sudden and comprehensive transformation. These claims, totally contradicting all the rules of genetics, biophysics, and biochemistry are as scientific as the fairy tales about frogs turning into princes! Nevertheless, being distressed by the crisis that the neo-Darwinist assertion was in, some evolutionist paleontologists embraced this theory, which had the distinction of being even more bizarre than neo-Darwinism itself.

The only purpose of this model was to provide an explanation of the gaps in the fossil-record that the neo-Darwinist model could not explain. However, it is hardly rational to attempt to explain the fossil gap in the evolution of birds with a claim that "**a bird popped all of a sudden out of a reptile egg**", because by the evolutionists' own admission, the evolution of a species to another species requires a great and advantageous change in genetic information. However, no mutation whatsoever improves the genetic information or adds new information to it. Mutations only derange genetic information. Thus the "gross mutations" imagined by the punctu-

Today, tens of thousands of scientists around the world, particularly in the USA and Europe, defy the theory of evolution and have published many books on the invalidity of the theory. Above are a few examples.

ated equilibrium model would only cause "gross", that is "great", reductions and impairments in the genetic information.

Moreover, the model of "punctuated equilibrium" collapses from the very first step by its inability to address the question of the origin of life, which is also the question that refutes the neo-Darwinist model from the outset. Since not even a single protein can have originated by chance, the debate over whether organisms made up of trillions of those proteins have undergone a "punctuated" or "gradual" evolution is senseless.

In spite of this, the model that comes to mind when "evolution" is at issue today is still neo-Darwinism. In the chapters that follow, we will first examine two imaginary mechanisms of the neo-Darwinist model and then look at the fossil record to test this model. After that, we will dwell upon the question of the origin of life, which invalidates both the neo-Darwinist model and all other evolutionist models such as "evolution by leaps".

Before doing so, it may be useful to remind the reader that the reality we will be confronting at every stage is that the evolutionary scenario is a fairy-tale, a great deceit that is totally at variance with the real world. It is a scenario that has been used to deceive the world for 140 years. Thanks to the latest scientific discoveries, its continued defence has at last become impossible.

CHAPTER 3

Imaginary Mechanisms of Evolution

The neo-Darwinist model, which we shall take as the mainstream theory of evolution today, argues that life has evolved through two natural mechanisms: "natural selection" and "mutation". The theory basically asserts that natural selection and mutation are two complementary mechanisms. The origin of evolutionary modifications lies in random mutations that take place in the genetic structures of living things. The traits brought about by mutations are selected by the mechanism of natural selection, and by this means living things evolve.

When we look further into this theory, we find that there is no such evolutionary mechanism. Neither natural selection nor mutations make any contribution at all to the transformation of different species into one another, and the claim that they do is completely unfounded.

Natural Selection

As process of nature, natural selection was familiar to biologists before Darwin, who defined it as a "mechanism that keeps species unchanging without being corrupted". Darwin was the first person to put forward the assertion that this process had evolutionary power and he then erected his entire theory on the foundation of this assertion. The name he gave to his book indicates that natural selection was the basis of Darwin's theory: *The Origin of Species, by means of Natural Selection...*

However since Darwin's time, there has not been a single shred of evidence put forward to show that natural selection causes living things to evolve. Colin Patterson, the senior paleontologist of the British Museum of Natural History in London and a prominent evolutionist, stresses that natural selection has never been observed to have the ability to cause things to evolve:

> **No one has ever produced a species by mechanisms of natural selection.** No one has ever got near it and most of the current argument in neo-Darwinism is about this question.[13]

Natural selection holds that those living things that are more suited to the natural conditions of their habitats will prevail by having offspring that will survive, whereas those that are unfit will disappear. For example, in a deer herd under the threat of wild animals, naturally those that can run faster will survive. That is true. But no matter how long this process goes on, it will not transform those deer into another living species. The deer will always remain deer.

When we look at the few incidents the evolutionists have put forth as observed examples of natural selection, we see that these are nothing but a simple attempt to hoodwink.

"Industrial Melanism"

In 1986 Douglas Futuyma published a book, *The Biology of Evolution*, which is accepted as one of the sources explaining the theory of evolution by natural selection in the most explicit way. The most famous of his examples on this subject is about the colour of the moth population, which appeared to darken during the Industrial Revolution in England. It is possible to find the story of the Industrial Melanism in almost all evolutionist biology books, not just in Futuyma's book. The story is based on a series of experiments conducted by the British physicist and biologist Bernard Kettlewell in the 1950s, and can be summarised as follows:

According to the account, around the onset of the Industrial Revolution in England, the colour of the tree barks around Manchester was quite light. Because of this, dark-coloured (melanic) moths resting on those trees could easily be noticed by the birds that fed on them and therefore they had very little chance of survival. Fifty years later, in woodlands where industrial pollution has killed the lichens, the barks of the trees had darkened, and now the light-colored moths became the most hunted, since they were the most easily noticed. As a result, the proportion of light-coloured moths to dark-coloured moths decreased. Evolutionists believe this to be a great piece of evidence for their theory. They take refuge and solace in window-dressing, showing how light-coloured moths "evolved" into dark-coloured ones.

However, although we believe these facts to be correct, it should be quite clear that they can in no way be used as evidence for the theory of evolution, since no new form arose that had not existed before. Dark colored moths had existed in the moth population before the Industrial Revo-

Industrial Melanism is certainly not an evidence for evolution because the process did not produce any new species of moths. The selection was only among already existing varieties. Moreover, the classical story of melanism is deceptive. The textbook pictures above (portrayed as genuine photos) are in fact of dead specimens glued or pinned to tree trunks by evolutionists.

lution. Only the relative proportions of the existing moth varieties in the population changed. The moths had not acquired a new trait or organ, which would cause "speciation". In order for one moth species to turn into another living species, a bird for example, new additions would have had to be made to its genes. That is, an entirely separate genetic program would have had to be loaded so as to include information about the physical traits of the bird.

This is the answer to be given to the evolutionist story of Industrial Melanism. However, there is a more interesting side to the story: Not just its interpretation, but the story itself is flawed. As molecular biologist Jonathan Wells explains in his book *Icons of Evolution*, the story of the peppered moths, which is included in every evolutionist biology book and has therefore, become an "icon" in this sense, does not reflect the truth. Wells discusses in his book how Bernard Kettlewell's experiment, which is known as the "experimental proof" of the story, is actually a scientific scandal. Some basic elements of this scandal are:

• Many experiments conducted after Kettlewell's revealed that only one type of these moths rested on tree trunks, and all other types preferred to rest beneath small, horizontal branches. Since 1980 it has become clear that peppered moths do not normally rest on tree trunks. In 25 years of fieldwork, many scientists such as Cyril Clarke and Rory Howlett, Michael Majerus, Tony Liebert, and Paul Brakefield concluded that "in Kettlewell's experiment, moths were forced to act atypically, therefore, the test results could not be accepted as scientific".

• Scientists who tested Kettlewell's conclusions came up with an even more interesting result: Although the number of light moths would be expected to be larger in the less polluted regions of England, the dark moths there numbered four times as many as the light ones. This meant that there was no correlation between the moth population and the tree trunks as claimed by Kettlewell and repeated by almost all evolutionist sources.

• As the research deepened, the scandal changed dimension: "The moths on tree trunks" photographed by Kettlewell, were actually dead moths. Kettlewell used dead specimens glued or pinned to tree trunks and then photographed them. In truth, there was little chance of taking such a picture as the moths rested not on tree trunks but underneath the leaves.[14]

These facts were uncovered by the scientific community only in the late 1990s. The collapse of the myth of Industrial Melanism, which had been one of the most treasured subjects in "Introduction to Evolution" courses in universities for decades, greatly disappointed evolutionists. One of them, Jerry Coyne, remarked:

> My own reaction resembles the dismay attending my discovery, at the age of six, that it was my father and not Santa who brought the presents on Christmas Eve.[15]

Thus, "the most famous example of natural selection" was relegated to the trash-heap of history as a scientific scandal which was inevitable, because natural selection is not an "evolutionary mechanism," contrary to what evolutionists claim. It is capable neither of adding a new organ to a living organism, nor of removing one, nor of changing an organism of one species into that of another.

Can Natural Selection Explain Complexity?

There is nothing that natural selection contributes to the theory of evolution, because this mechanism can **never increase or improve the ge-**

Natural selection serves as a mechanism of eliminating weak individuals within a species. It is a conservative force which preserves the existing species from degeneration. Beyond that, it has no capability of transforming one species to another.

netic information of a species. Neither can it transform one species into another: a starfish into a fish, a fish into a frog, a frog into a crocodile, or a crocodile into a bird. The biggest defender of punctuated equilibrium, Stephen Jay Gould, refers to this impasse of natural selection as follows;

> The essence of Darwinism lies in a single phrase: natural selection is the creative force of evolutionary change. No one denies that selection will play a negative role in eliminating the unfit. Darwinian theories require that it create the fit as well.[16]

Another of the misleading methods that evolutionists employ on the issue of natural selection is their effort to present this mechanism as a conscious designer. However, **natural selection has no consciousness**. It does not possess a will that can decide what is good and what is bad for living things. As a result, natural selection cannot explain biological systems and organs that possess the feature of "**irreducible complexity**". These systems and organs are composed of a great number of parts cooperating together, and are of no use if even one of these parts is missing or defective. (For example, the human eye does not function unless it exists with all its components intact). Therefore, the will that brings all these parts together should be able to foresee the future and aim directly at the advantage that is to be acquired at the final stage. Since natural selection has no consciousness or will, it can do no such thing. This fact, which demolishes the foundations of the theory of evolution, also worried Darwin, who wrote: "**If it could be demonstrated that any complex organ existed, which could not possibly have been formed by numerous, successive, slight modifications, my theory would absolutely break down.**"[17]

Natural selection only selects out the disfigured, weak, or unfit individuals of a species. It cannot produce new species, new genetic information, or new organs. That is, it cannot make anything evolve. Darwin accepted this reality by saying: "**Natural selection can do nothing until favourable variations chance to occur**".[18] This is why neo-Darwinism has had to elevate mutations next to natural selection as the "cause of beneficial changes". However as we shall see, mutations can only be "the cause for harmful changes".

Mutations

Mutations are defined as breaks or replacements taking place in the DNA molecule, which is found in the nuclei of the cells of a living organism and which contains all its genetic information. These breaks or replacements are the result of external effects such as radiation or chemical action. Every mutation is an "accident" and either damages the nucleotides making up the DNA or changes their locations. Most of the time, they cause so much damage and modification that the cell cannot repair them.

Mutation, which evolutionists frequently hide behind, is not a magic wand that transforms living organisms into a more advanced and perfect form. The direct effect of mutations is harmful. The changes effected by mutations can only be like those experienced by people in Hiroshima, Nagasaki, and Chernobyl: that is, death, disability, and freaks of nature...

The reason for this is very simple: DNA has a very complex structure, and random effects can only damage it. B.G. Ranganathan states:

> First, genuine mutations are very rare in nature. Secondly, most mutations are harmful since they are random, rather than orderly changes in the structure of genes; any random change in a highly ordered system will be for the worse, not for the better. For example, if an earthquake were to shake a highly ordered structure such as a building, there would be a random change in the framework of the building which, in all probability, would not be an improvement. [19]

Not surprisingly, **no useful mutation has been so far observed**. All mutations have proved to be harmful. The evolutionist scientist Warren Weaver comments on the report prepared by the Committee on Genetic Effects of Atomic Radiation, which had been formed to investigate mutations that might have been caused by the nuclear weapons used in the Second World War:

ALL MUTATIONS ARE HARMFUL

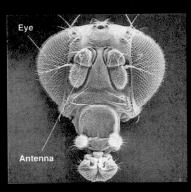
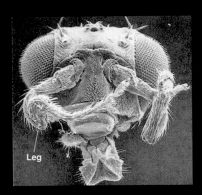
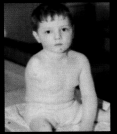

Left: A normal fruit fly (drosophila).
Right: A fruit fly with its legs jutting from its head; a mutation induced by radiation.

A disastrous effect of mutations on the human body. The boy at left is a Chernobyl nuclear plant accident victim.

Many will be puzzled about the statement that practically all known mutant genes are harmful. For mutations are a necessary part of the process of evolution. How can a good effect - evolution to higher forms of life - results from **mutations practically all of which are harmful**?[20]

Every effort put into "generating a useful mutation" has resulted in failure. For decades, evolutionists carried out many experiments to produce mutations in **fruit flies** as these insects reproduce very rapidly and so mutations would show up quickly. Generation upon generation of these flies were mutated, yet no useful mutation was ever observed. The evolutionist geneticist Gordon Taylor writes thus:

> **It is a striking, but not much mentioned fact that, though geneticists have been breeding fruit-flies for sixty years or more in labs all around the world-flies which produce a new generation every eleven days-they have never yet seen the emergence of a new species or even a new enzyme.**[21]

Another researcher, Michael Pitman, comments on the failure of the experiments carried out on fruit flies:

> Morgan, Goldschmidt, Muller, and other geneticists have subjected generations of fruit flies to extreme conditions of heat, cold, light, dark, and treatment by chemicals and radiation. All sorts of mutations, practically all trivial

or positively deleterious, have been produced. Man-made evolution? Not really: Few of the geneticists' monsters could have survived outside the bottles they were bred in. In practice **mutants die, are sterile, or tend to revert to the wild type**.[22]

The same holds true for man. All mutations that have been observed in human beings have had deleterious results. On this issue, evolutionists throw up a smokescreen and try to enlist examples of even such deleterious mutations as "evidence for evolution". All mutations that take place in humans result in physical deformities, in infirmities such as **mongolism, Down syndrome, albinism, dwarfism or cancer**. These mutations are presented in evolutionist textbooks as examples of "the evolutionary mechanism at work". Needless to say, a process that leaves people disabled or sick cannot be "an evolutionary mechanism"-evolution is supposed to produce forms that are better fitted to survive.

To summarise, there are three main reasons why mutations cannot be pressed into the service of supporting evolutionists' assertions:

1) The direct effect of mutations is harmful: Since they occur randomly, they almost always damage the living organism that undergoes them. Reason tells us that unconscious intervention in a perfect and complex structure will not improve that structure, but will rather impair it. Indeed, no "useful mutation" has ever been observed.

2) Mutations add no new information to an organism's DNA: The particles making up the genetic information are either torn from their places, destroyed, or carried off to different places. Mutations cannot make a living thing acquire a new organ or a new trait. They only cause abnormalities like a leg sticking out of the back, or an ear from the abdomen.

3) In order for a mutation to be transferred to the subsequent generation, it has to have taken place in the reproductive cells of the organism: A random change that occurs in a cell or organ of the body cannot be transferred to the next generation. For example, a human eye altered by the effects of radiation or by other causes will not be passed on to subsequent generations.

Briefly, it is impossible for living beings to have evolved, because there exists no mechanism in nature that can cause evolution. Furthermore, this conclusion agrees with the evidence of the fossil record, which does not demonstrate the existence of a process of evolution, but rather just the contrary.

CHAPTER 4

The Fossil Record Refutes Evolution

According to the theory of evolution, every living species has emerged from a predecessor. One species which existed previously turned into something else over time and all species have come into being in this way. According to the theory, this transformation proceeds gradually over millions of years.

If this were the case, then innumerable intermediate species should have lived during the immense period of time when these transformations were supposedly occurring. For instance, there should have lived in the past some half-fish/half-reptile creatures which had acquired some reptilian traits in addition to the fish traits they already had. Or there should have existed some reptile/bird creatures, which had acquired some avian traits in addition to the reptilian traits they already possessed. Evolutionists refer to these imaginary creatures, which they believe to have lived in the past, as "transitional forms".

If such animals had really existed, there would have been millions, even billions, of them. More importantly, the remains of these creatures should be present in the fossil record. The number of these transitional forms should have been even greater than that of present animal species, and their remains should be found all over the world. In *The Origin of Species*, Darwin accepted this fact and explained:

> If my theory be true, numberless intermediate varieties, linking most closely all of the species of the same group together must assuredly have existed... Consequently evidence of their former existence could be found only amongst fossil remains.[23]

Even Darwin himself was aware of the absence of such transitional forms. He hoped that they would be found in the future. Despite his optimism, he realised that these missing intermediate forms were the biggest stumbling-block for his theory. That is why he wrote the following in the chapter of the *The Origin of Species* entitled "Difficulties of the Theory":

> ...Why, if species have descended from other species by fine gradations, **do we not everywhere see innumerable transitional forms?** Why is not all na-

ture in confusion, instead of the species being, as we see them, well defined?... But, as by this theory innumerable transitional forms must have existed, why do we not find them embedded in countless numbers in the crust of the earth?... But in the intermediate region, having intermediate conditions of life, why do we not now find closely-linking intermediate varieties? This difficulty for a long time quite confounded me.[24]

The only explanation Darwin could come up with to counter this objection was the argument that the fossil record uncovered so far was inadequate. He asserted that when the fossil record had been studied in detail, the missing links would be found.

Believing in Darwin's prophecy, evolutionist paleontologists have been digging up fossils and searching for missing links all over the world since the middle of the 19th century. Despite their best efforts, **no transitional forms have yet been uncovered**. All the fossils unearthed in excavations have shown that, contrary to the beliefs of evolutionists, life appeared on earth all of a sudden and fully-formed. Trying to prove their theory, evolutionists have instead unwittingly caused it to collapse.

A famous British paleontologist, Derek V. Ager, admits this fact even though he is an evolutionist:

> The point emerges that if we examine the fossil record in detail, whether at the level of orders or of species, we find-over and over again-**not gradual evolution, but the sudden explosion of one group at the expense of another.**[25]

Another evolutionist paleontologist Mark Czarnecki comments as follows:

> A major problem in proving the theory has been the fossil record; the imprints of vanished species preserved in the Earth's geological formations. This record has never revealed traces of Darwin's hypothetical intermediate variants - **instead species appear and disappear abruptly**, and this anomaly has fueled the creationist argument that each species was created by God.[26]

These gaps in the fossil record cannot be explained by saying that sufficient fossils have not yet been found, but that they one day will be. Another American scholar, Robert Wesson, states in his 1991 book *Beyond Natural Selection*, that "the gaps in the fossil record are real and meaningful". He elaborates this claim in this way:

> The gaps in the record are real, however. The absence of a record of any important branching is quite phenomenal. Species are usually static, or nearly so, for long periods, species seldom and genera never show evolution into

Living Fossils

The theory of evolution claims that species continuously evolve into other species. But when we compare living things to their fossils, we see that they have remained unchanged for millions of years. This fact is a clear evidence that falsifies the claims of evolutionists.

The living honeybee is no different than its fossil relative, which is millions of years old.

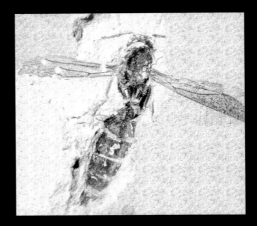

The 135 million year old dragon fly fossil is no different than its modern counterparts.

A comparison of ant fossil aged 100 million years and an ant living in our day clearly indicates that ants do not have any evolutionary history.

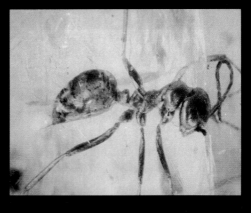

new species or genera but replacement of one by another, and change is more or less abrupt.[27]

Life Emerged on Earth Suddenly and in Complex Forms

When terrestrial strata and the fossil record are examined, it is to be seen that all living organisms appeared simultaneously. The oldest stratum of the earth in which fossils of living creatures have been found is that of the Cambrian, which has an estimated age of 500-550 million years.

The living creatures found in the strata belonging to the Cambrian period emerged all of a sudden in the fossil record-there are no pre-existing ancestors. The fossils found in Cambrian rocks belonged to snails, trilobites, sponges, earthworms, jellyfish, sea hedgehogs, and other complex invertebrates. This wide mosaic of living organisms made up of such a great number of complex creatures emerged so suddenly that this miraculous event is referred to as the "Cambrian Explosion" in geological literature.

Most of the creatures in this layer have complex systems have complex systems and advanced structures, such as eyes, gills, and circulatory systems, exactly the same as those in modern specimens. For instance, the double-lensed, combed eye structure of trilobites is a wonder of design. David Raup, a professor of geology in Harvard, Rochester, and Chicago Universities, says: **"the trilobites 450 million years ago used an optimal design which would require a well trained and imaginative optical engineer to develop today"**.[28]

These complex invertebrates emerged suddenly and completely without having any link or any transitional form between them and the unicellular organisms, which were the only life forms on earth prior to them.

Richard Monastersky, a science journalist at *Science News*, one of the popular publications of evolutionist literature, states the following about the "Cambrian Explosion", which is a deathtrap for evolutionary theory:

> A half-billion years ago, the remarkably complex forms of animals we see today suddenly appeared. This moment, right at the start of Earth's Cambrian Period, some 550 million years ago, marks the evolutionary explosion that filled the seas with the earth's first complex creatures. ...the large animal phyla of today were present already in the early Cambrian ...and they were as distinct from each other as they are today.[29]

Deeper investigation into the Cambrian Explosion shows what a

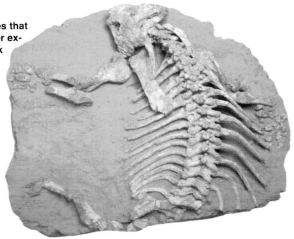

The fossil record proves that transitional forms never existed, no evolution took place and all species have been created separately in a perfect form.

great dilemma it creates for the theory of evolution. Recent findings indicate that almost all phyla, the most basic animal divisions, emerged abruptly in the Cambrian period. An article published in *Science* magazine in 2001 says: "The beginning of the Cambrian period, some 545 million years ago, saw the sudden appearance in the fossil record of almost all the main types of animals (phyla) that still dominate the biota today".[30] The same article notes that for such complex and distinct living groups to be explained according to the theory of evolution, very rich fossil beds showing a gradual developmental process should have been found, but this has not yet proved possible:

> This differential evolution and dispersal, too, must have required a previous history of the group for which there is no fossil record. Furthermore, cladistic analyses of arthropod phylogeny revealed that trilobites, like eucrustaceans, are fairly advanced "twigs" on the arthropod tree. But fossils of these alleged ancestral arthropods are lacking. ...Even if evidence for an earlier origin is discovered, it remains a challenge to explain why so many animals should have increased in size and acquired shells within so short a time at the base of the Cambrian.[31]

How the earth came to overflow with such a great number of animal species all of a sudden, and how these distinct types of species with no common ancestors could have emerged, is a question that remains unanswered by evolutionists. The Oxford University zoologist Richard Dawkins, one of the foremost advocates of evolutionist thought in the world, comments on this reality that undermines the very foundation of all the arguments he has been defending:

For example the Cambrian strata of rocks... are the oldest ones in which we find most of the major invertebrate groups. And we find many of them already in an advanced state of evolution, the very first time they appear. **It is as though they were just planted there, without any evolutionary history.**[32]

As Dawkins is forced to acknowledge, the Cambrian Explosion is strong evidence for creation, because creation is the only way to explain the fully-formed emergence of life on earth. Douglas Futuyma, a prominent evolutionist biologist admits this fact: "Organisms either appeared on the earth fully developed or they did not. If they did not, they must have developed from pre-existing species by some process of modification. If **they did appear in a fully developed state, they must indeed have been created by some omnipotent intelligence."** [33] Darwin himself recognised the possibility of this when he wrote: "If numerous species, belonging to the same genera or families, have really started into life all at once, **the fact would be fatal to the theory of descent with slow modification through natural selection.**"[34] The Cambrian Period is nothing more or less than

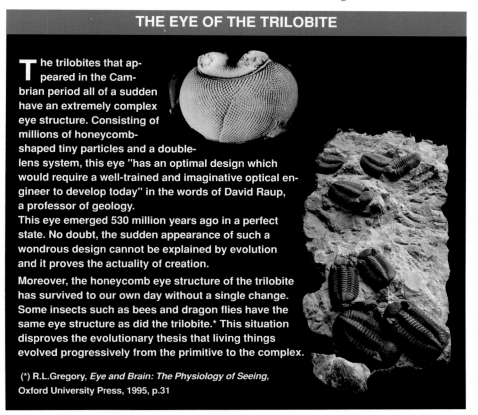

THE EYE OF THE TRILOBITE

The trilobites that appeared in the Cambrian period all of a sudden have an extremely complex eye structure. Consisting of millions of honeycomb-shaped tiny particles and a double-lens system, this eye "has an optimal design which would require a well-trained and imaginative optical engineer to develop today" in the words of David Raup, a professor of geology.

This eye emerged 530 million years ago in a perfect state. No doubt, the sudden appearance of such a wondrous design cannot be explained by evolution and it proves the actuality of creation.

Moreover, the honeycomb eye structure of the trilobite has survived to our own day without a single change. Some insects such as bees and dragon flies have the same eye structure as did the trilobite.* This situation disproves the evolutionary thesis that living things evolved progressively from the primitive to the complex.

(*) R.L.Gregory, *Eye and Brain: The Physiology of Seeing*, Oxford University Press, 1995, p.31

Darwin's "fatal stroke". This is why the Swiss evolutionist paleoanthropologist Stefan Bengtson, who confesses the lack of transitional links while describing the Cambrian Age, makes the following comment: "Baffling (and embarrassing) to Darwin, this event still dazzles us".[35]

As may be seen, the fossil record indicates that living things did not evolve from primitive to the advanced forms, but instead emerged all of a sudden and in a perfect state. In short, living beings did not come into existence by evolution, they were created.

Molecular Comparisons Deepen Evolution's Cambrian Impasse

Another fact that puts evolutionists into a deep quandary about the Cambrian Explosion is the comparisons between different living taxa. The results of these comparisons reveal that animal taxa considered to be "close relatives" by evolutionists until quite recently, are genetically very different, which puts the "intermediate form" hypothesis, that only exists theoretically, into an even greater quandary. An article published in the Proceedings of the National Academy of Sciences in 2000 reports that DNA analyses have displaced taxa that used to be considered "intermediate forms" in the past:

> DNA sequence analysis dictates new interpretation of phylogenic trees. Taxa that were once thought to represent successive grades of complexity at the base of the metazoan tree are being displaced to much higher positions inside the tree. This leaves no evolutionary "intermediates" and forces us to rethink the genesis of bilaterian complexity...[36]

In the same article, evolutionist writers note that some taxa which were considered "intermediate" between groups such as sponges, cnidarians and ctenophores can no longer be considered as such because of new genetic findings, and that they have "lost hope" of constructing such evolutionary family trees:

> The new molecular based phylogeny has several important implications. Foremost among them is the disappearance of "intermediate" taxa between sponges, cnidarians, ctenophores, and the last common ancestor of bilaterians or "Urbilateria." ...A corollary is that we have a major gap in the stem leading to the Urbilataria. We have lost the hope, so common in older evolutionary reasoning, of reconstructing the morphology of the "coelomate ancestor" through a scenario involving successive grades of increasing complexity based on the anatomy of extant "primitive" lineages.[37]

CHAPTER 5

Tale of Transition from Water to Land

Evolutionists assume that the sea invertebrates that appear in the Cambrian stratum somehow evolved into fish in tens of million years. However, just as Cambrian invertebrates have no ancestors, there are no transitional links indicating that an evolution occurred between these invertebrates and fish. It should be noted that invertebrates and fish have enormous structural differences. Invertebrates have their hard tissues outside their bodies, whereas fish are vertebrates that have theirs on the inside. Such an enormous "evolution" would have taken billions of steps to be completed and there should be billions of transitional forms displaying them.

Evolutionists have been digging fossil strata for about 140 years looking for these hypothetical forms. They have found millions of invertebrate fossils and millions of fish fossils; yet nobody has ever found even one that is midway between them.

An evolutionist paleontologist, Gerald T. Todd, admits a similar fact in an article titled "Evolution of the Lung and the Origin of Bony Fishes":

> All three subdivisions of bony fishes first appear in the fossil record at approximately the same time. They are already widely divergent morphologically, and are heavily armored. How did they originate? What allowed them to diverge so widely? How did they all come to have heavy armour? And why is there no trace of earlier, intermediate forms?[38]

According to the hypothetical scenario of "from sea to land", some fish felt the need to pass from sea to land because of feeding problems. This claim is "supported" by such speculative drawings.

The evolutionary scenario goes one step further and argues that fish, who evolved from invertebrates then transformed into amphibians. But this scenario also lacks evidence. There is not even a single fossil verifying that a half-fish/half-amphibian creature has ever existed. Robert L. Carroll, an evolutionary palaeontologist and authority on vertebrate palaeontology, is obliged to accept this. He has written in his classic work, *Vertebrate Paleontology and Evolution*, that "The early reptiles were very different from amphibians and their ancestors have not been found yet." In his newer book, *Patterns and Processes of Vertebrate Evolution*, published in 1997, he admits that "The origin of the modern amphibian orders, (and) the transition between early tetrapods" are "still poorly known" along with the origins of many other major groups.[39] Two evolutionist paleontologists, Colbert and Morales, comment on the three basic classes of amphibians- frogs, salamanders, and caecilians:

> **There is no evidence of any Paleozoic amphibians combining the characteristics that would be expected in a single common ancestor.** The oldest known frogs, salamanders, and caecilians are very similar to their living descendants.[40]

Until about fifty years ago, evolutionists thought that such a creature indeed existed. This fish, called a coelacanth, which was estimated to be 410 million years of age, was put forward as a transitional form with a primitive lung, a developed brain, a digestive and a circulatory system ready to function on land, and even a primitive walking mechanism. These

410-million-year-old coelacanth fossil. Evolutionists claimed that it was the transitional form representing the transition from water to land. Living examples of this fish have been caught many times since 1938, providing a good example of the extent of the speculations that evolutionists engage in.

Why Transition From Water to Land is Impossible

FOCUS

Evolutionists claim that one day, a species dwelling in water somehow stepped onto land and was transformed into a land-dwelling species. There are a number of obvious facts that render such a transition impossible:

1. Weight-bearing: Sea-dwelling creatures have no problem in bearing their own weight in the sea.
However, most land-dwelling creatures consume 40% of their energy just in carrying their bodies around. Creatures making the transition from water to land would at the same time have had to develop new muscular and skeletal systems (!) to meet this energy need, and this could not have come about by chance mutations.

2. Heat Retention: On land, the temperature can change quickly, and fluctuates over a wide range. Land-dwelling creatures possess a physical mechanism that can withstand such great temperature changes. However, in the sea, the temperature changes slowly and within a narrower range. A living organism with a body system regulated according to the constant temperature of the sea would need to acquire a protective system to ensure minimum harm from the temperature changes on land. It is preposterous to claim that fish acquired such a system by random mutations as soon as they stepped onto land.

3. Water: Essential to metabolism, water needs to be used economically due to its relative scarcity on land. For instance,, the skin has to be able to permit a certain amount of water loss, while also preventing excessive evaporation. That is why land-dwelling creatures experience thirst, something the land-dwelling creatures do not do. For this reason, the skin of sea-dwelling animals is not suitable for a nonaquatic habitat.

4. Kidneys: Sea-dwelling organisms discharge waste materials, especially ammonia, by means of their aquatic environment. On land, water has to be used economically. This is why these living beings have a kidney system. Thanks to the kidneys, ammonia is stored by being converted into urea and the minimum amount of water is used during its excretion. In addition, new systems are needed to provide the kidney's functioning. In short, in order for the passage from water to land to have occurred, living things without a kidney would have had to develop a kidney system all at once.

5. Respiratory system: Fish "breathe" by taking in oxygen dissolved in water that they pass through their gills. They canot live more than a few minutes out of water. In order to survive on land, they would have to acquire a perfect lung system all of a sudden.

It is most certainly impossible that all these dramatic physiological changes could have happened in the same organism at the same time, and all by chance.

TURTLES WERE ALWAYS TURTLES

Turtle fossil aged 100 million years: No different from its modern counterpart. (*The Dawn of Life*, Orbis Pub., London 1972)

Just as the evolutionary theory cannot explain basic classes of living things such as fish and reptiles, neither can it explain the origin of the orders within these classes. For example, turtles, which is a reptilian order, appear in the fossil record all of a sudden with their unique shells. To quote from an evolutionary source: "...by the middle of the Triassic Period (about 175,000,000 years ago) its (turtle's) members were already numerous and in possession of the basic turtle characteristics. The links between turtles and cotylosaurs from which turtles probably sprang are almost entirely lacking" (*Encyclopaedia Brittanica*, 1971, v.22, p.418).

There is no difference between the fossils of ancient turtles and the living members of this species today. Simply put, turtles have not "evolved"; they have always been

anatomical interpretations were accepted as undisputed truth among scientific circles until the end of the 1930's. The coelacanth was presented as a genuine transitional form that proved the evolutionary transition from water to land.

However on December 22, 1938, a very interesting discovery was made in the Indian Ocean. A living member of the coelacanth family, previously presented as a transitional form that had become extinct seventy million years ago, was caught! The discovery of a "living" prototype of the coelacanth undoubtedly gave evolutionists a severe shock. The evolutionist paleontologist J.L.B. Smith said that "If I'd met a dinosaur in the street I wouldn't have been more astonished".[41] In the years to come, 200 coelacanths were caught many times in different parts of the world.

Living coelacanths revealed how far the evolutionists could go in making up their imaginary scenarios. Contrary to what had been claimed, coelacanths had neither a primitive lung nor a large brain. The organ that evolutionist researchers had proposed as a primitive lung turned out to be nothing but a lipid pouch.[42] Furthermore, the coelacanth, which was introduced as "a reptile candidate getting prepared to pass from sea to land", was in reality a fish that lived in the depths of the oceans and never approached nearer than 180 metres from the surface.[43]

CHAPTER 6

Origin of Birds and Mammals

According to the theory of evolution, life originated and evolved in the sea and then was transported onto land by amphibians. This evolutionary scenario also suggests that amphibians evolved into reptiles, creatures living only on land. This scenario is again implausible, due to the enormous structural differences between these two classes of animals. For instance, the amphibian egg is designed for developing in water whereas the amniotic egg is designed for developing on land. A "step by step" evolution of an amphibian is out of the question, because without a perfect and fully-designed egg, it is not possible for a species to survive. Moreover, as usual, there is no evidence of transitional forms that were supposed to link amphibians with reptiles. Evolutionist paleontologist and an authority on vertebrate paleontology, Robert L. Carroll has to accept that **"the early reptiles were very different from amphibians and that their ancestors could not be found yet."**[44]

Yet the hopelessly doomed scenarios of the evolutionists are not over yet. There still remains the problem of making these creatures fly! Since evolutionists believe that birds must somehow have been evolved, they assert that they were transformed from reptiles. However, none of the distinct mechanisms of birds, which have a completely different structure from land-dwelling animals, can be explained by gradual evolution. First of all, the wings, which are the exceptional traits of birds, are a great impasse for the evolutionists. One of the Turkish evolutionists, Engin Korur, confesses the impossibility of the evolution of wings:

> The common trait of the eyes and the wings is that they can only function if they are fully developed. In other words, **a halfway-developed eye cannot see; a bird with half-formed wings cannot fly.** How these organs came into being has remained one of the mysteries of nature that needs to be enlightened.[45]

The question of how the perfect structure of wings came into being as a result of consecutive haphazard mutations remains completely unanswered. There is no way to explain how the front arms of a reptile could have changed into perfectly functioning wings as a result of a distortion in its genes (mutation).

Moreover, just having wings is not sufficient for a land organism to fly. Land-dwelling organisms are devoid of many other structural mechanisms that birds use for flying. For example, the bones of birds are much lighter than those of land-dwelling organisms. Their lungs function in a very different way. They have a different muscular and skeletal system and a very specialised heart-circulatory system. These features are pre-requisites of flying needed at least as much as wings. All these mechanisms had to exist at the same time and altogether; they could not have formed gradually by being "accumulated". This is why the theory asserting that land organisms evolved into aerial organisms is completely fallacious.

All of these bring another question to the mind: even if we suppose this impossible story to be true, then why are the evolutionists unable to find any "half-winged" or "single-winged" fossils to back up their story?

Another Alleged Transitional Form: *Archæopteryx*

Evolutionists pronounce the name of one single creature in response. This is the fossil of a bird called *Archæopteryx*, one of the most widely-known so-called transitional forms among the very few that evolutionists

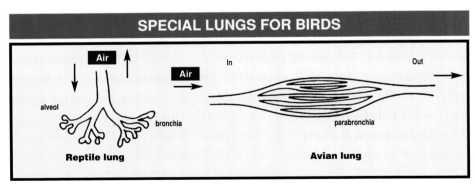

The anatomy of birds is very different from that of reptiles, their supposed ancestors. Bird lungs function in a totally different way from those of land-dwelling animals. Land-dwelling animals breathe in and out from the same air vessel. In birds, while the air enters into the lung from the front, it goes out from the back. This distinct "design" is specially made for birds, which need great amounts of oxygen during flight. It is impossible for such a structure to evolve from the reptile lung.

still defend. *Archæopteryx*, the so-called ancestor of modern birds according to evolutionists, lived approximately 150 million years ago. The theory holds that some small dinosaurs, such as *Velociraptors* or *Dromeosaurs*, evolved by acquiring wings and then starting to fly. Thus, *Archæopteryx* is assumed to be a transitional form that branched off from its dinosaur ancestors and started to fly for the first time.

However, the latest studies of *Archæopteryx* fossils indicate that this creature is absolutely not a transitional form, but an extinct species of bird, having some insignificant differences from modern birds.

The thesis that *Archæopteryx* was a "half-bird" that could not fly perfectly was popular among evolutionist circles until not long ago. The absence of a sternum (breastbone) in this creature was held up as the most important evidence that this bird could not fly properly. (The sternum is a bone found under the thorax to which the muscles required for flight are attached. In our day, this breastbone is observed in all flying and non-flying birds, and even in bats, a flying mammal which belongs to a very different family.)

However, **the seventh *Archæopteryx* fossil, which was found in 1992,** caused great astonishment among evolutionists. The reason was that in this recently discovered fossil, the breastbone that was long assumed by evolutionists to be missing was discovered to have existed after all. This fossil was described in *Nature* magazine as follows:

> The recently discovered seventh specimen of the Archaeopteryx preserves a partial, rectangular sternum, long suspected but never previously documented. **This attests to its strong flight muscles**.[46]

This discovery invalidated the mainstay of the claims that *Archæopteryx* was a half-bird that could not fly properly.

Moreover, the structure of the bird's feathers became one of the most important pieces of evidence confirming that *Archæopteryx* was a flying bird in the real sense. The asymmetric feather structure of *Archæopteryx* is indistinguishable from that of modern birds, and indicates that it could fly perfectly well. As the eminent paleontologist Carl O. Dunbar states, "because of its feathers [*Archæopteryx* is] distinctly to be classed as a bird."[47]

Another fact that was revealed by the structure of *Archæopteryx*'s feathers was its warm-blooded metabolism. As was discussed above, reptiles and dinosaurs are cold-blooded animals whose body heat fluctuates with the temperature of their environment, rather than being homeostatically regulated. A very important function of the feathers on birds is the maintenance of a constant body temperature. The fact that *Archæopteryx* had feathers showed that it was a real, warm-blooded bird that needed to regulate its body heat, in contrast to dinosaurs.

Speculations of Evolutionists:
The Teeth and Claws of *Archæopteryx*

Two important points evolutionist biologists rely on when claiming *Archæopteryx* was a transitional form, are the claws on its wings and its teeth.

It is true that *Archæopteryx* had claws on its wings and teeth in its mouth, but these traits do not imply that the creature bore any kind of relationship to reptiles. Besides, two bird species living today, Taouraco and Hoatzin, have claws which allow them to hold onto branches. These creatures are fully birds, with no reptilian characteristics. That is why it is completely groundless to assert that *Archæopteryx* is a transitional form just because of the claws on its wings.

Neither do the teeth in *Archæopteryx*'s beak imply that it is a transitional form. Evolutionists make a purposeful trickery by saying that these teeth are reptile characteristics, since teeth are not a typical feature of reptiles. Today, some reptiles have teeth while others do not. Moreover, *Archæopteryx* is not the only bird species to possess teeth. It is true that there are no toothed birds in existence today, but when we look at the fossil record, we see that both during the time of *Archæopteryx* and afterwards, and even until fairly recently, a distinct bird genus existed that could be categorised as "birds with teeth".

The most important point is that the **tooth structure of *Archæopteryx* and other birds with teeth is totally different from that of** their alleged ancestors, **the dinosaurs.** The well-known ornithologists L. D. Martin, J. D. Steward, and K. N. Whetstone observed that *Archæopteryx* and other similar birds have teeth with flat-topped surfaces and large roots. Yet the teeth of theropod dinosaurs, the alleged ancestors of these birds, are protuberant like saws and have narrow roots.[48]

These researchers also compared the wrist bones of *Archæopteryx* and their alleged ancestors, the dinosaurs, and observed no similarity between them.[49]

Studies by anatomists like S. Tarsitano, M. K. Hecht, and A.D. Walker have revealed that some of the similarities that John Ostrom and other have seen between *Archæopteryx* and dinosaurs were in reality misinterpretations.[50]

All these findings indicate that *Archæopteryx* was not a transitional link but only a bird that fell into a category that can be called "toothed birds".

Archæopteryx and Other Ancient Bird Fossils

While evolutionists have for decades been proclaiming *Archæopteryx* to be the greatest evidence for their scenario concerning the evolution of birds, some recently-found fossils invalidate that scenario in other respects.

Lianhai Hou and Zhonghe Zhou, two paleontologists at the Chinese Institute of Vertebrate Paleontology, discovered a new bird fossil in 1995, and named it ***Confuciusornis***. This fossil is almost the same age as *Archæopteryx* (around 140 million years), but has no teeth in its mouth. In addition, its beak and feathers shared the same features as today's birds. *Confuciusornis* has the same skeletal structure as modern birds, but also has claws on its wings, just like *Archæopteryx*. Another structure peculiar to birds called the "pygostyle", which supports the tail feathers, was also found in *Confuciusornis*. In short, this fossil-which is the same age as *Archæopteryx*, which was previously thought to be the earliest bird and was accepted as a semi-reptile-looks very much like a modern bird. This fact has invalidated all the evolutionist theses claiming *Archæopteryx* to be the primitive ancestor of all birds.[51]

Another fossil unearthed in China, caused even greater confusion. In November 1996,

When bird feathers are examined in detail, it is seen that they are made up of thousands of tiny tendrils attached to one another with hooks. This unique design results in superior aerodynamic performance.

The Design of the Bird Feathers

The theory of evolution, which claims that birds evolved from reptiles, is unable to explain the huge differences between these two different living classes. In terms of such features as their skeleton structure, lung systems, and warm-blooded metabolism, birds are very different from reptiles. Another trait that poses an insurmountable gap between birds and reptiles is the feathers of birds which have a form entirely peculiar to them.

The bodies of reptiles are covered with scales, whereas the bodies of birds are covered with feathers. Since evolutionists consider reptiles the ancestor of birds, they are obliged to claim that bird feathers have evolved from reptile scales. However, there is no similarity between scales and feathers.

A professor of physiology and neurobiology from the University of Connecticut, A.H. Brush, accepts this reality although he is an evolutionist: "Every feature from gene structure and organization, to development, morphogenesis and tissue organization is different (in feathers and scales)."[1] Moreover, Prof. Brush examines the protein structure of bird feathers and argues that it is "unique among vertebrates".[2]

There is no fossil evidence to prove that bird feathers evolved from reptile scales. On the contrary, "feathers appear suddenly in the fossil record, as an 'undeniably unique' character distinguishing birds" as Prof. Brush states.[3] Besides, in reptiles, no epidermal structure has yet been detected that provides an origin for bird feathers.[4]

In 1996, paleontologists made a buzz about fossils of a so-called feathered dinosaur, called Sinosauropteryx. However in 1997, it was revealed that these fossils had nothing to do with birds and that they were not modern feathers.[5]

On the other hand, when we examine bird feathers closely, we come across a very complex design that cannot be explained by any evolutionary process. The famous ornithologist Alan Feduccia states that "every feature of them has aerodynamic functions. They are extremely light, have the ability to lift up which increases in lower speeds, and may return to their previous position very easily". Then he continues, "I cannot really understand how an organ perfectly designed for flight may have emerged for another need at the beginning".[6]

The design of feathers also compelled Charles Darwin ponder them. Moreover, the perfect aesthetics of the peafowl's feathers had made him "sick" (his own words). In a letter he wrote to Asa Gray on April 3, 1860, he said "I remember well the time when the thought of the eye made me cold all over, but I have got over this stage of complaint..." And then continued: "...and now trifling particulars of structure often make me very uncomfortable. The sight of a feather in a peacock's tail, whenever I gaze at it, makes me sick!"[7]

1- A. H. Brush, "On the Origin of Feathers". *Journal of Evolutionary Biology*, Vol. 9, 1996, p.132
2- A. H. Brush, "On the Origin of Feathers". p. 131
3- *Ibid.*
4- *Ibid.*
5- "Plucking the Feathered Dinosaur", *Science*, Vol. 278, 14 November 1997, p. 1229
6- Douglas Palmer, "Learning to Fly" (*Review of The Origin of and Evolution of Birds* by Alan Feduccia, Yale University Press, 1996), *New Scientist*, Vol. 153, March, 1 1997, p. 44
7- Norman Macbeth, *Darwin Retried: An Appeal to Reason*. Boston, Gambit, 1971, p. 101

the existence of a 130-million-year-old bird named **Liaoningornis** was announced in *Science* by L. Hou, L. D. Martin, and Alan Feduccia. *Liaoningornis* had a breastbone to which the muscles for flight were attached, just as in modern birds. This bird was indistinguishable from modern birds also in other respects, too. The only difference was the teeth in its mouth. This showed that birds with teeth did not possess the primitive structure alleged by evolutionists.[52] This was stated in an article in *Discover* "Whence came the birds? This fossil suggests that it was not from dinosaur stock".[53]

Another fossil that refuted the evolutionist claims regarding *Archæopteryx* was *Eoalulavis*. The wing structure of **Eoalulavis**, which was said to be some 25 to 30 million years younger than *Archæopteryx*, was also observed in modern slow-flying birds. This proved that 120 million years ago, there were birds indistinguishable from modern birds in many respects flying in the skies.[54]

These facts once more indicate for certain that neither *Archæopteryx* nor other ancient birds similar to it were transitional forms. The fossils do not indicate that different bird species evolved from each other. On the contrary, the fossil record proves that today's modern birds and some archaic birds such as *Archæopteryx* actually lived together at the same time. It is true that some of these bird species, such as *Archæopteryx* and *Confuciusornis*, have become extinct, but the fact that only some of the species that once existed have been able to survive down to the present day does not in itself support the theory of evolution.

In brief, several features of *Archæopteryx* indicate that this creature was not a transitional form. The overall anatomy of *Archæopteryx* imply stasis, not evolution. Paleontologist Robert Carroll has to admit that:

> The geometry of the flight feathers of Archaeopteryx is identical with that of modern flying birds, whereas nonflying birds have symmetrical feathers. The way in which the feathers are arranged on the wing also falls within the range of modern birds... According to Van Tyne and Berger, the relative size and shape of the wing of Archaeopteryx are similar to that of birds that move through restricted openings in vegetation, such as gallinaceous birds, doves, woodcocks, woodpeckers, and most passerine birds... The flight feathers have been in stasis for at least 150 million years...[55]

The Imaginary Bird-Dinosaur Link

The claim of evolutionists trying to present *Archæopteryx* as a transitional form is that birds have evolved from dinosaurs. However, one of the most famous ornithologists in the world, Alan Feduccia from the University of North Carolina, opposes the theory that birds are related to dinosaurs, despite the fact that he is an evolutionist himself. Feduccia has this to say regarding the thesis of reptile-bird evolution:

Prof. Alan Feduccia

> Well, I've studied bird skulls for 25 years and I don't see any similarities whatsoever. I just don't see it... The theropod origins of birds, in my opinion, will be the greatest embarrassment of paleontology of the 20th century.[56]

Larry Martin, a specialist on ancient birds from the University of Kansas, also opposes the theory that birds are descended from dinosaurs. Discussing the contradiction that evolution falls into on the subject, he states:

> To tell you the truth, if I had to support the dinosaur origin of birds with those characters, I'd be embarrassed every time I had to get up and talk about it.[57]

To sum up, the scenario of the "evolution of birds" erected solely on the basis of *Archæopteryx*, is nothing more than a product of the prejudices and wishful thinking of evolutionists.

The bird named *Confuciusornis* is the same age as *Archæopteryx*

What is the Origin of Flies?

An example from evolutionist scenarios: Dinosaurs that suddenly took wing while trying to catch flies!

Claiming that dinosaurs transformed into birds, evolutionists support their assertion by saying that some dinosaurs who flapped their front legs to hunt flies "took wing and flew" as seen in the picture. Having no scientific basis whatsoever and being nothing but a figment of the imagination, this theory also entails a very simple logical contradiction: the example given by evolutionists to explain the origin of flying, that is, the fly, already has a perfect ability to fly. Whereas a human cannot open and close his eyes 10 times a second, an average fly flutters its wings 500 times a second. Moreover, it moves both its wings simultaneously. The slightest dissonance in the vibration of wings would cause the fly lose its balance but this never happens.

Evolutionists should first come up with an explanation as to how the fly acquired this perfect ability to fly. Instead, they fabricate imaginary scenarios about how much more clumsy creatures like reptiles came to fly.

Even the perfect creation of the housefly invalidates the claim of evolution. English biologist Robin Wootton wrote in an article titled "The Mechanical Design of Fly Wings":

The better we understand the functioning of insect wings, the more subtle and beautiful their designs appear. Structures are traditionally designed to deform as little as possible; mechanisms are designed to move component parts in predictable ways. Insect wings combine both in one, using components with a wide range of elastic properties, elegantly assembled to allow appropriate deformations in response to appropriate forces and to make the best possible use of the air. They have few if any technological parallels-yet.[1]

On the other hand, there is not a single fossil that can be evidence for the imaginary evolution of flies. This is what the distinguished French zoologist Pierre Grassé meant when he said "We are in the dark concerning the origin of insects."[2]

1- Robin J. Wootton, "The Mechanical Design of Insect Wings", *Scientific American*, v. 263, November 1990, p.120
2- Pierre-P Grassé, *Evolution of Living Organisms*, New York, Academic Press, 1977, p.30

The Origin of Mammals

As we have stated before, the theory of evolution proposes that some imaginary creatures that came out of the sea turned into reptiles, and that birds evolved from reptiles. According to the same scenario, reptiles are the ancestors not only of birds but also of mammals. However, there are great differences between these two classes. Mammals are warm-blooded animals (this means they can generate their own heat and maintain it at a steady level), they give live birth, they suckle their young, and their bodies are covered in fur or hair. Reptiles, on the other hand, are cold-blooded (i.e., they cannot generate heat, and their body temperature changes according to the external temperature), they lay eggs, they do not suckle their young, and their bodies are covered in scales.

One example of the structural barriers between reptiles and mammals is their **jaw structure**. Mammal jaws consist of only one mandibular bone containing the teeth. In reptiles, there are three little bones on both sides of the mandible. Another basic difference is that all mammals have three bones in their middle ear (hammer, anvil, and stirrup). Reptiles have but a single bone in the middle ear. Evolutionists claim that the reptile jaw and middle ear gradually evolved into the mammal jaw and ear. The question of how an ear with a single bone evolved into one with three bones, and how the sense of hearing kept on functioning in the meantime can never be explained. Not surprisingly, not one single fossil linking reptiles and mammals has been found. This is why evolutionist science writer Roger Lewin was forced to say, "**The transition to the first mammal, which probably happened in just one or, at most, two lineages, is still an enigma**".[58]

George Gaylord Simpson, one of the most popular evolutionary authorities and a founder of the neo-Darwinist theory, makes the following comment regarding this perplexing difficulty for evolutionists:

> The most puzzling event in the history of life on earth is **the change from the Mesozoic, the Age of Reptiles, to the Age of Mammals**. It is as if the curtain were rung down suddenly on the stage where all the leading roles were taken by reptiles, especially dinosaurs, in great numbers and bewildering variety, and rose again immediately to reveal the same setting but an entirely new cast, a cast in which the dinosaurs do not appear at all, other reptiles are supernumeraries, and **all the leading parts are played by mammals of sorts barely hinted at in the preceding acts**.[59]

Furthermore, when mammals suddenly made their appearance, they were already very different from each other. Such dissimilar animals as **bats, horses, mice, and whales** are all mammals, and they all emerged during the same geological period. Establishing an evolutionary relationship among them is impossible even by the broadest stretch of the imagination. The evolutionist zoologist R. Eric Lombard makes this point in an article that appeared in the leading journal *Evolution*:

> Those searching for specific information useful in constructing phylogenies of mammalian taxa will be disappointed.[60]

All of these demonstrate that all living beings appeared on earth suddenly and fully formed, without any evolutionary process. This is concrete evidence of the fact that they were created. Evolutionists, however, try to interpret the fact that living species came into existence in a particular order as an indication of evolution. Yet the sequence by which living things emerged is the "**order of creation**", since it is not possible to speak of an evolutionary process. With a superior and flawless creation, oceans and then lands were filled with living things and finally man was created.

Evolutionists propose that all mammal species evolved from a common ancestor. However, there are great differences between various mammal species such as bears, whales, mice, and bats. Each of these living beings possesses specifically designed systems. For example, bats are created with a very sensitive sonar system that helps them find their way in darkness. These complex systems, which modern technology can only imitate, could not possibly have emerged as a result of chance coincidence. The fossil record also demonstrates that bats came into being in their present perfect state all of a sudden and that they have not undergone any "evolutionary process".

A bat fossil aged 50 million years: no different from its modern counterpart. (*Science*, vol. 154)

Contrary to the "ape man" story that is imposed on the masses with intense media propaganda, man also emerged on earth suddenly and fully formed.

The Myth of Horse Evolution

Until recently, an imaginary sequence supposedly showing the evolution of the horse was advanced as the principal fossil evidence for the theory of evolution. Today, however, many evolutionists themselves frankly admit that the scenario of horse evolution is bankrupt. In 1980, a four-day symposium was held at the Field Museum of Natural History in Chicago, with 150 evolutionists in attendance, to discuss the problems with gradualistic evolutionary theory. In addressing the meeting, evolutionist Boyce Rensberger noted that the scenario of the evolution of the horse has no foundation in the fossil record, and that no evolutionary process has been observed that would account for the gradual evolution of horses:

> The popularly told example of horse evolution, suggesting a gradual sequence of changes from four-toed fox-sized creatures living nearly 50 million years ago to today's much larger one-toed horse, has long been known to be wrong. Instead of gradual change, fossils of each intermediate species appear fully distinct, persist unchanged, and then become extinct. Transitional forms are unknown.[1]

The well-known paleontologist Colin Patterson, a director of the Natural History Museum in London where "evolution of the horse" diagrams were on public display at that time on the ground floor of the museum, said the following about the exhibition:

> There have been an awful lot of stories, some more imaginative than others, about what the nature of that history [of life] really is. The most famous example, still on exhibit downstairs, is the exhibit on horse evolution prepared perhaps fifty years ago. That has been presented as the literal truth in textbook after textbook. Now I think that is lamentable, particularly when the people who propose those kinds of stories may themselves be aware of the speculative nature of some of that stuff.[2]

Then what is the basis for the scenario of the evolution of the horse? This scenario was formulated by means of the deceitful charts devised by the sequential arrangement of fossils of distinct species that lived at vastly different periods in India, South Africa, North America, and Europe solely in accordance with the rich power of evolutionists' imaginations. More than 20 charts of the evolution of the horse, which by the way are totally different from each other, have been proposed by various researchers. Thus, it is obvious that evolutionists have reached no common agreement on these family trees. The only common feature in these arrangements is the belief that a dog-sized creature called "*Eohippus*", which lived in the Eocene Period 55 million years ago, was the ancestor of the horse (*Equus*). But, the supposed evolutionary lines from *Eohippus* to *Equus* are totally inconsistent.

The evolutionist science writer Gordon R. Taylor explains this little-acknowledged truth in his book *The Great Evolution Mystery*:

> But perhaps the most serious weakness of Darwinism is the failure of paleontologists to find convincing phylogenies or sequences of organisms demonstrating major evolutionary change... The horse is often cited as the only fully worked-out example. But the fact is that the line from *Eohippus* to *Equus* is very erratic. It is alleged to show a continual increase in size, but the truth is that some variants were smaller than *Eohippus*, not larger. Specimens from different sources can be brought together in a convincing-looking sequence, but there is no evidence that they were actually ranged in this order in time.[3]

All these facts are strong evidence that the charts of horse evolution, which are presented as one of the most solid pieces of evidence for Darwinism, are nothing but fantastic and implausible tales.

1- Boyce Rensberger, *Houston Chronicle*, November 5, 1980, p.15
2- Colin Patterson, *Harper's*, February 1984, p.60
3- Gordon Rattray Taylor, *The Great Evolution Mystery*, Abacus, Sphere Books, London, 1984, p. 230

CHAPTER 7

Deceptive Fossil Interpretations

Before going into the details of the myth of human evolution, we need to mention the propaganda method that has convinced the general public of the idea that half-man half-ape creatures once lived in the past. This propaganda method makes use of "reconstructions" made in reference to fossils. Reconstruction can be explained as drawing a picture or constructing a model of a living thing based on a single bone-sometimes only a fragment-that has been unearthed. The "ape-men" we see in newspapers, magazines, or films are all reconstructions.

Since fossils are usually fragmented and incomplete, any conjecture based on them is likely to be completely speculative. As a matter of fact, the reconstructions (drawings or models) made by the evolutionists based on fossil remains are prepared speculatively precisely to validate the evolutionary thesis. David R. Pilbeam, an eminent anthropologist from Harvard, stresses this fact when he says: "At least in paleoanthropology, data are still so sparse that theory heavily influences interpretations. **Theories have, in the past, clearly reflected our current ideologies instead of the actual data**".[61] Since people are highly affected by visual information, these reconstructions best serve the purpose of evolutionists, which is to convince people that these reconstructed creatures really existed in the past.

At this point, we have to highlight one particular point: Reconstructions based on bone remains can only reveal the most general characteristics of the creature, since the really distinctive morphological features of any animal are soft tissues which quickly vanish after death. Therefore, due to the speculative nature of the interpretation of the soft tissues, the reconstructed drawings or models become totally dependent on the imagination of the person producing them. Earnst A. Hooten from Harvard University explains the situation like this:

> To attempt to restore the soft parts is an even more hazardous undertaking. The lips, the eyes, the ears, and the nasal tip leave no clues on the underlying bony parts. **You can with equal facility model on a Neanderthaloid skull**

Imaginary and Deceptive Drawings

In pictures and reconstructions, evolutionists deliberately give shape to features that do not actually leave any fossil traces, such as the structure of the nose and lips, the shape of the hair, the form of the eyebrows, and other bodily hair so as to support evolution. They also prepare detailed pictures depicting these imaginary creatures walking with their families, hunting, or in other instances of their daily lives. However, these drawings are all figments of the imagination and have no counterpart in the fossil record.

THREE DIFFERENT RECONSTRUCTIONS BASED ON THE SAME SKULL

Appeared in *Sunday Times*
April 5, 1964

Maurice Wilson's
drawing

N.Parker's reconstruction
N. Geographic, September 1960

the features of a chimpanzee or the lineaments of a philosopher. These alleged restorations of ancient types of man have very little if any scientific value **and are likely only to mislead the public... So put not your trust in reconstructions.**[62]

As a matter of fact, evolutionists invent such "preposterous stories" that they even ascribe different faces to the same skull. For example, the three different reconstructed drawings made for the fossil named *Australopithecus robustus* **(Zinjanthropus)**, are a famous example of such forgery.

The biased interpretation of fossils and outright fabrication of many imaginary reconstructions are an indication of how frequently evolutionists have recourse to tricks. Yet these seem innocent when compared to the deliberate forgeries that have been perpetrated in the history of evolution.

CHAPTER 8

Evolution Forgeries

T here is no concrete fossil evidence to support the "ape-man" image, which is unceasingly promulgated by the media and evolutionist academic circles. With brushes in their hands, evolutionists produce imaginary creatures, nevertheless, the fact that these drawings correspond to no matching fossils constitutes a serious problem for them. One of the interesting methods they employ to overcome this problem is to **"produce" the fossils they cannot find.** Piltdown Man, which may be the biggest scandal in the history of science, is a typical example of this method.

Piltdown Man: An Orang-utan Jaw and a Human Skull!

In 1912, a well-known doctor and amateur paleoanthropologist named Charles Dawson came out with the assertion that he had found a jawbone and a cranial fragment in a pit in Piltdown, England. Even though the jawbone was more ape-like, the teeth and the skull were like a man's. These specimens were labelled the "Piltdown man". Alleged to be 500,000 years old, they were displayed as an absolute proof of human evolution in several museums. For more than 40 years, many scientific articles were written on "Piltdown man", many interpretations and drawings were made, and the fossil was presented as important evidence for human evolution. No fewer than 500 doctoral theses were written on the subject.[63] While visiting the British Museum in 1921, leading American paleoanthropologist Henry Fairfield Osborn said "We have to be reminded over and over again that Nature is full of paradoxes" and proclaimed Piltdown "a discovery of transcendant importance to the prehistory of man".[64]

In 1949, Kenneth Oakley from the British Museum's Paleontology Department, attempted to use "fluorine testing", a new test used for determining the date of fossils. A trial was made on the fossil of the Piltdown man. The result was astonishing. During the test, it was realised that the

The Story of a Hoax

1 The fossils are unearthed by Charles Dawson and given to Sir Arthur Smith Woodward.

2 Pieces are reconstructed to form the famous skull.

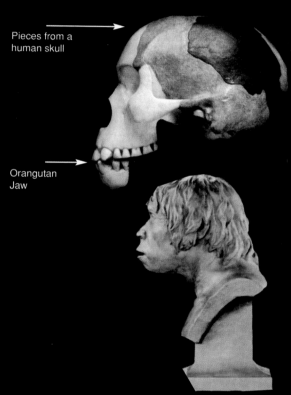

Pieces from a human skull

Orangutan Jaw

3 Based on the reconstructed skull, various drawings and skulptures are made, numerous articles and commentaries are written. The original skull is demonstrated in the British Museum.

4 After 40 years of its discovery, the Piltdown fossil is shown to be a hoax by a group of researchers.

jawbone of Piltdown Man did not contain any fluorine. This indicated that it had remained buried no more than a few years. The skull, which contained only a small amount of fluorine, showed that it was not older than a few thousand years old.

It was determined that the teeth in the jawbone belonging to an orangutan, had been worn down artificially and that the "primitive" tools discovered with the fossils were simple imitations that had been sharpened with steel implements.[65] In the detailed analysis completed by Joseph Weiner, this forgery was revealed to the public in 1953. **The skull belonged to a 500-year-old man, and the jaw bone belonged to a recently deceased ape!** The teeth had been specially arranged in a particular way and added to the jaw, and the molar surfaces were filed in order to resemble those of a man. Then all these pieces were stained with potassium dichromate to give them an old appearance. These stains began to disappear when dipped in acid. Sir Wilfred Le Gros Clark, who was in the team that uncovered the forgery, could not hide his astonishment at this situation and said: "**The evidences of artificial abrasion immediately sprang to the eye**. Indeed so obvious did they seem it may well be asked-how was it that they had escaped notice before?"[66] In the wake of all this, "Piltdown man" was hurriedly removed from the British Museum where it had been displayed for more than 40 years.

Nebraska Man: A Pig's Tooth

In 1922, Henry Fairfield Osborn, the director of the American Museum of Natural History, declared that he had found a fossil molar tooth belonging to the Pliocene period in western Nebraska near Snake Brook. This tooth allegedly bore common characteristics of both man and ape. An extensive scientific debate began surrounding this fossil, which came to be called "Nebraska man", in which some interpreted this tooth as belonging to *Pithecanthropus erectus*, while others claimed it was closer to human beings. Nebraska man was also immediately given a "scientific name", *Hesperopithecus haroldcooki*.

Many authorities gave Osborn their support. **Based on this single tooth, reconstructions of the Nebraska man's head and body were drawn.** Moreover, Nebraska man was even pictured along with his wife and children, as a whole family in a natural setting.

Evolution Forgeries

The picture on the left was drawn on the basis of a single tooth and it was published in the *Illustrated London News* magazine on July 24, 1922. However, the evolutionists were extremely disappointed when it was revealed that this tooth belonged neither to an ape-like creature nor to a man, but rather to an extinct pig species.

All of these scenarios were developed from just one tooth. Evolutionist circles placed such faith in this "ghost man" that when a researcher named William Bryan opposed these biased conclusions relying on a single tooth, he was harshly criticised.

In 1927, other parts of the skeleton were also found. According to these newly discovered pieces, the tooth belonged neither to a man nor to an ape. It was realised that it belonged to an extinct species of wild American pig called *Prosthennops*. William Gregory entitled the article published in *Science* in which he announced the truth, "*Hesperopithecus*: Apparently Not an ape Nor a man".[67] Then all the drawings of *Hesperopithecus haroldcooki* and his "family" were hurriedly removed from evolutionary literature.

Ota Benga: The African In The Cage

After Darwin advanced the claim with his book *The Descent of Man* that man evolved from ape-like living beings, he started to seek fossils to support this contention. However, some evolutionists believed that "half-man half-ape" creatures were to be found not only in the fossil record, but also alive in various parts of the world. In the early 20th century, these pursuits for **"living transitional links"** led to unfortunate incidents, one of the cruellest of which is the story of a Pygmy by the name of Ota Benga.

Ota Benga was captured in 1904 by an evolutionist researcher in the Congo. In his own tongue, his name meant "friend". He had a wife and two children. Chained and caged like an animal, he was taken to the USA where evolutionist scientists displayed him to the public in the St Louis World Fair along with other ape species and introduced him as "**the closest** transitional link to man". Two years later, they took him to the Bronx Zoo in New York and there they exhibited him under the denomination of "ancient ancestors of man" along with a few chimpanzees, a gorilla named Dinah, and an orang-utan called Dohung. Dr William T. Hornaday, the zoo's evolutionist director gave long speeches on how proud he was to have this exceptional "transitional form" in his zoo and treated caged Ota Benga as if he were an ordinary animal. Unable to bear the treatment he was subjected to, Ota Benga eventually committed suicide.[68]

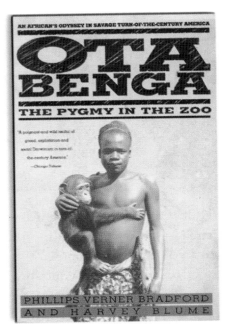

OTA BENGA:
"The pygmy in the zoo"

Piltdown Man, Nebraska Man, Ota Benga... These scandals demonstrate that evolutionist scientists do not hesitate to employ any kind of unscientific method to prove their theory. Bearing this point in mind, when we look at the other so-called evidence of the "human evolution" myth, we confront a similar situation. Here there are a fictional story and an army of volunteers ready to try everything to verify this story.

CHAPTER 9

The Scenario of Human Evolution

In previous chapters, we saw that there are no mechanisms in nature to lead the living beings to evolve and that living species came into existence not as the result of an evolutionary process, but rather emerged all of a sudden in their present perfect structure. That is, they were created individually. Therefore, it is obvious that "human evolution", too, is a story that has never taken place.

What, then, do the evolutionists propose as the basis for this story?

This basis is the existence of plenty of fossils on which the evolutionists are able to build up imaginary interpretations. Throughout history, more than **6,000** ape species have lived and most of them have become extinct. Today, only **120** ape species live on the earth. These approximately 6,000 ape species, most of which are extinct, constitute a rich resource for the evolutionists.

The evolutionists wrote the scenario of human evolution by arranging some of the skulls that suited their purpose in an order from the smallest to the biggest and scattering the skulls of some extinct human races among them. According to this scenario, men and modern apes have common ancestors. These creatures evolved in time and some of them became the apes of today while another group that followed another branch of evolution became the men of today.

However, all the paleontological, anatomical and biological findings have demonstrated that this claim of evolution is as fictitious and invalid as all the others. No sound or real evidence has been put forward to prove that there is a relationship between man and ape, except forgeries, distortions, and misleading drawings and comments.

The fossil record indicates to us that throughout history, men have been men and apes have been apes. Some of the fossils the evolutionists claim to be the ancestors of man, belong to human races that lived until very recently-about 10,000 years ago-and then disappeared. Moreover, many human communities currently living have the same physical appearance and characteristics as these extinct human races, which the evolutionists claim to be the ancestors of men. All these are clear proof that man has never gone through an evolutionary process at any period in history.

The most important of all is that there are numerous anatomical differences between apes and men and none of them are of the kind to come into existence through an evolutionary process. "**Bipedality**" is one of them. As we will describe later on in detail, bipedality is peculiar to man and it is one of the most important traits that distinguishes man from other animals.

The Imaginary Family Tree of Man

The Darwinist claim holds that modern man evolved from some kind of ape-like creature. During this alleged evolutionary process, which is supposed to have started from 4 to 5 million years ago, it is claimed that there existed some "transitional forms" between modern man and his ancestors. According to this completely imaginary scenario, the following four basic "categories" are listed:

1. Australopithecines (any of the various forms belonging to the genus *Australopithecus*)

2. *Homo habilis*

3. *Homo erectus*

4. *Homo sapiens*

Evolutionists call the genus to which the alleged ape-like ancestors of man belonged "***Australopithecus***", which means "southern ape". *Australopithecus*, which is nothing but an old type of ape that has become extinct, is found in various different forms. Some of them are larger and strongly built (robust), while others are smaller and delicate (gracile).

Evolutionists classify the next stage of human evolution as the genus ***Homo***, that is "man". According to the evolutionist claim, the living things in the *Homo* series are more developed than *Australopithecus*, and not very much different from modern man. The modern man of our day, that is, the species *Homo sapiens*, is said to have formed at the latest stage of the evolution of this genus *Homo*.

Fossils like "**Java Man**", "**Pekin Man**", and "**Lucy**", which appear in the media from time to time and are to be found in evolutionist publications and textbooks, are included in one of the four groups listed above.

A SINGLE JAWBONE AS A SPARK OF INSPIRATION

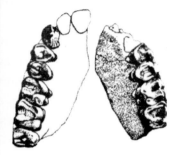

The first Ramapithecus fossil found: a missing jaw composed of two parts. (on the right). The evolutionists daringly pictured Ramapithecus, his family and the environment they lived in, by relying only on these jawbones.

Each of these groupings is also assumed to branch into species and subspecies, as the case may be.

Some suggested transitional forms of the past, such as **Ramapithecus**, had to be excluded from the imaginary human family tree after it was realised that they were ordinary apes.[69]

By outlining the links in the chain as "australopithecines > *Homo habilis* > *Homo erectus* > *Homo sapiens*", the evolutionists imply that each of these types is the ancestor of the next. However, recent findings by paleoanthropologists have revealed that australopithecines, *Homo habilis* and *Homo erectus* existed in different parts of the world at the same time. Moreover, some of those humans classified as *Homo erectus* probably lived up until very modern times. In an article titled "Latest *Homo erectus* of Java: Potential Contemporaneity with *Homo sapiens* in Southeast Asia", it was reported in the journal *Science* that *Homo erectus* fossils found in Java had "mean ages of 27 ± 2 to 53.3 ± 4 thousand years ago" and this "raise[s] the possibility that *H. erectus* overlapped in time with anatomically modern humans (*H. sapiens*) in Southeast Asia"[70]

Furthermore, *Homo sapiens neandarthalensis* and *Homo sapiens sapiens* (modern man) also clearly co-existed. This situation apparently indicates the invalidity of the claim that one is the ancestor of the other.

Intrinsically, all findings and scientific research have revealed that the fossil record does not suggest an evolutionary process as evolutionists propose. The fossils, which evolutionists claim to be the ancestors of humans, in fact belong either to different human races, or else to species of ape.

Then which fossils are human and which ones are apes? Is it ever possible for any one of them to be considered a transitional form? In order to find the answers, let us have a closer look at each category.

Australopithecus: An Ape Species

The first category, the genus *Australopithecus*, means "southern ape", as we have said. It is assumed that these creatures first appeared in Africa about 4 million years ago, and lived until 1 million years ago. There are a number of different species among the astralopithecines. Evolutionists assume that the oldest *Australopithecus* species is *A. Afarensis*. After that comes *A. Africanus*, and then *A. Robustus*, which has relatively bigger bones. As for *A. Boisei*, some researchers accept it as a different species, and others as a sub-species of *A. Robustus*.

All of the *Australopithecus* species are extinct apes that resemble the apes of today. Their cranial capacities are the same or smaller than the chimpanzees of our day. There are projecting parts in their hands and feet which they used to climb trees, just like today's chimpanzees, and their feet are built for grasping to hold onto branches. They are short (maximum 130 cm. (51 in.)) and just like today's chimpanzees, male *Australopithecus* is larger than the female. Many other characteristics-such as the details in their skulls, the closeness of their eyes, their sharp molar teeth, their mandibular structure, their long arms, and their short legs-constitute evidence that these creatures were no different from today's ape.

However, evolutionists claim that, although australopithecines have the anatomy of apes, unlike apes, they walked upright like humans.

This claim that australopithecines **walked upright** is a view that has been held by paleoanthropologists such as Richard Leakey and Donald C. Johanson for decades. Yet many scientists who have carried out a great deal of research on the skeletal structures of australopithecines have proved the invalidity of that argument. Extensive research done on various *Australopithecus* specimens by two world-renowned anatomists from England and the USA, Lord Solly Zuckerman and Prof. Charles Oxnard, showed that these creatures did not walk upright in human manner. Having studied the bones of these fossils for a period of 15 years thanks to grants from the British government, Lord Zuckerman and his team of five specialists reached the conclusion that australopithecines were only **an ordinary ape genus** and were **definitely not bipedal**, although Zuckerman is an evolutionist himself.[71] Correspondingly, Charles E. Oxnard, who is another evolutionist famous for his research on the subject, also likened the skeletal structure of australopithecines to that of modern orang-utans.[72]

Briefly, Australopithecines have no link with humans and they are merely an extinct ape species.

Homo Habilis: The Ape that was Presented as Human

The great similarity between the skeletal and cranial structures of australopithecines and chimpanzees, and the refutation of the claim that these creatures walked upright, have caused great difficulty for evolutionist paleoanthropologists. The reason is that, according to the imaginary evolution scheme, *Homo erectus* comes after *Australopithecus*. As the genus name *Homo* (meaning "man") implies, **Homo erectus** is a human species and its skeleton is straight. Its cranial capacity is twice as large as that of *Australopithecus*. A direct transition from *Australopithecus*, which is a chimpanzee-like ape, to *Homo erectus*, which has a skeleton no different from modern man's, is out of the question even according to evolutionist theory. Therefore, "links"-that is, "transitional forms"-are needed. The concept of **Homo habilis** arose from this necessity.

The classification of *Homo habilis* was put forward in the 1960s by the Leakeys, a family of "fossil hunters". According to the Leakeys, this new species, which they classified as *Homo habilis*, had a relatively large cranial capacity, the ability to walk upright and to use stone and wooden tools. Therefore, it could have been the ancestor of man.

New fossils of the same species unearthed in the late 1980s, were to completely change this view. Some researchers, such as Bernard Wood and C. Loring Brace, who relied on those newly-found fossils, stated that *Homo*

Australopithecus Aferensis: An Extinct Ape

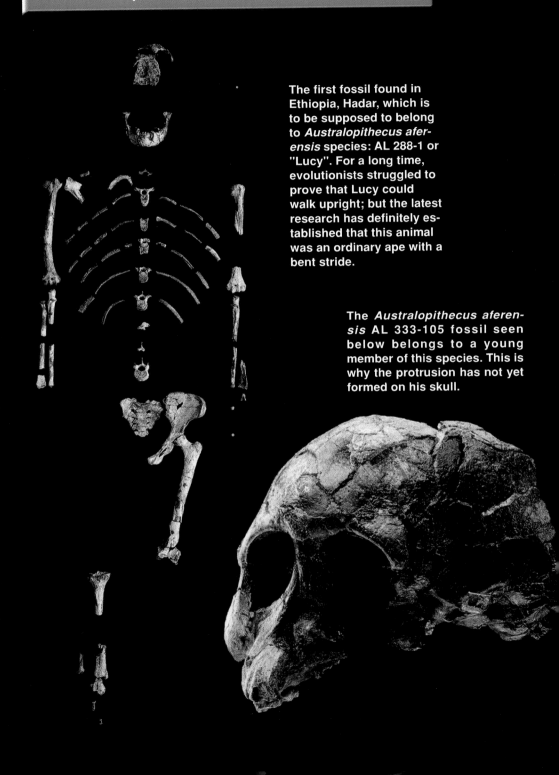

The first fossil found in Ethiopia, Hadar, which is to be supposed to belong to *Australopithecus aferensis* species: AL 288-1 or "Lucy". For a long time, evolutionists struggled to prove that Lucy could walk upright; but the latest research has definitely established that this animal was an ordinary ape with a bent stride.

The *Australopithecus aferensis* AL 333-105 fossil seen below belongs to a young member of this species. This is why the protrusion has not yet formed on his skull.

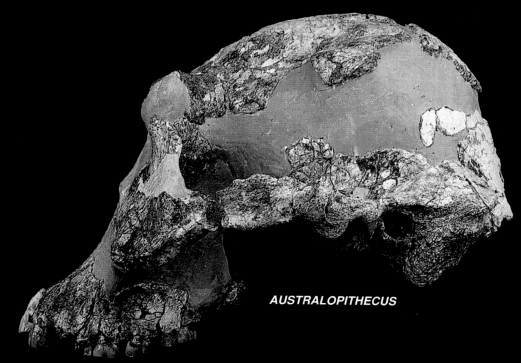

AUSTRALOPITHECUS

Above is seen the skull of *Australopithecus* aferensis AL 444-2 fossil, and below is the skull of a contemporary ape. The obvious similarity verifies that *A. aferensis* is an ordinary ape species without any "human-like" features.

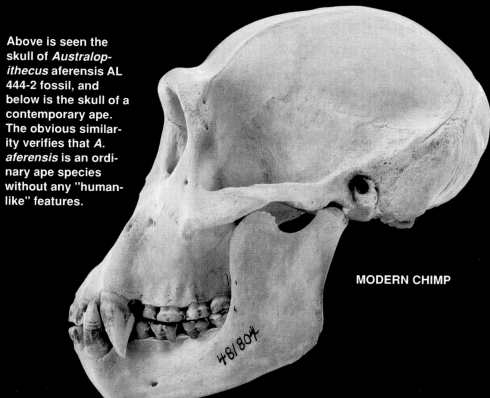

MODERN CHIMP

habilis (which means "skillful man"; that is, man capable of using tools) should be classified as *Australopithecus habilis*, or "skillful southern ape", because *Homo habilis* had a lot of characteristics in common with the australopithecine apes. It had long arms, short legs and an ape-like skeletal structure just like *Australopithecus*. Its fingers and toes were suitable for climbing. Their jaw was very similar to that of today's apes. Their 600 cc average cranial capacity is also an indication of the fact that they were apes. In short, *Homo habilis*, which was presented as a different species by some evolutionists, was in reality an ape species just like all the other australopithecines.

Research carried out in the years since Wood and Brace's work has demonstrated that *Homo habilis* was indeed no different from *Australopithecus*. The skull and skeletal fossil OH62 found by Tim White showed that this species had a **small cranial capacity**, as well as **long arms** and **short legs** which enabled them to climb trees just like modern apes do.

The detailed analyses conducted by American anthropologist Holly Smith in 1994 indicated that *Homo habilis* was not *Homo*, in other words, "human", at all, but rather unequivocally an "ape". Speaking of the analyses she made on the teeth of *Australopithecus*, *Homo habilis*, *Homo erectus* and *Homo neanderthalensis*, Smith stated the following;

> Restricting analysis of fossils to specimens satisfying these criteria, **patterns of dental development of gracile australopithecines and *Homo Habilis* remain classified with African apes**. Those of *Homo erectus* and Neanderthals are classified with humans.[73]

Within the same year, Fred Spoor, Bernard Wood and Frans Zonneveld, all specialists on anatomy, reached a similar conclusion through a totally different method. This method was based on the comparative analysis of the semi-circular canals in the inner ear of humans and apes which provided for sustaining balance. Spoor, Wood and Zonneveld concluded that:

> Among the fossil hominids the earliest species to demonstrate the modern human morphology is *Homo erectus*. In contrast, the semi-circular canal dimensions in crania from southern Africa attributed to *Australopithecus* and Paranthropus resemble those of the extant great apes.[74]

Spoor, Wood and Zonneveld also studied a *Homo habilis* specimen, namely Stw 53, and found out that "Stw 53 relied less on bipedal behavior than the australopithecines." This meant that the *H. habilis* specimen was even more ape-like than the *Australopithecus* species. Thus they concluded

Homo Habilis: Another Extinct Ape

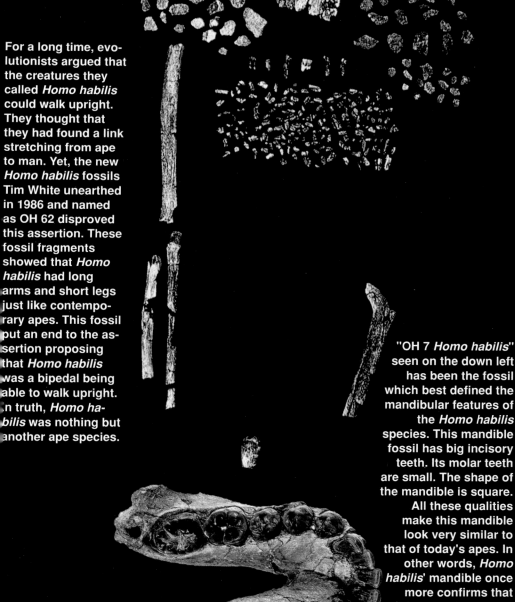

For a long time, evolutionists argued that the creatures they called *Homo habilis* could walk upright. They thought that they had found a link stretching from ape to man. Yet, the new *Homo habilis* fossils Tim White unearthed in 1986 and named as OH 62 disproved this assertion. These fossil fragments showed that *Homo habilis* had long arms and short legs just like contemporary apes. This fossil put an end to the assertion proposing that *Homo habilis* was a bipedal being able to walk upright. In truth, *Homo habilis* was nothing but another ape species.

"OH 7 *Homo habilis*" seen on the down left has been the fossil which best defined the mandibular features of the *Homo habilis* species. This mandible fossil has big incisory teeth. Its molar teeth are small. The shape of the mandible is square. All these qualities make this mandible look very similar to that of today's apes. In other words, *Homo habilis*' mandible once more confirms that this being is actually an ape.

that "Stw 53 represents an unlikely intermediate between the morphologies seen in the australopithecines and *H. erectus*."

This finding yielded two important results:

1. Fossils referred to as *Homo habilis* did not actually belong to the genus *Homo*, i.e. humans, but to that of *Australopithecus*, i.e. apes.

2. Both *Homo habilis* and *Australopithecus* were creatures that walked stooped forward-that is to say, they had the skeleton of an ape. They have no relation whatsoever to man.

Homo Rudolfensis: The Face Wrongly Joined

The term *Homo rudolfensis* is the name given to a few fossil fragments unearthed in 1972. The species supposedly represented by this fossil was designated *Homo rudolfensis* because these fossil fragments were found in the vicinity of Lake Rudolf in Kenya. Most of the paleoanthropologists accept that these fossils do not belong to a distinct species, but that the creature called *Homo rudolfensis* is in fact indistinguishable from *Homo habilis*.

Richard Leakey, who unearthed the fossils, presented the skull designated "KNM-ER 1470", which he said was 2.8 million years old, as the greatest discovery in the history of anthropology. According to Leakey, this creature, which had a small cranial capacity like that of *Australopithecus* together with a face similar to that of present-day humans, was the missing link between *Australopithecus* and humans. Yet, after a short while, it was realised that the human-like face of the KNM-ER 1470 skull, which frequently appeared on the covers of scientific journals and popular science magazines was the result of the incorrect assembly of the skull fragments, which may have been deliberate. Professor Tim Bromage, who conducts studies on human facial anatomy, brought this to light by the help of computer simulations in 1992:

> When it [KNM-ER 1470] was first reconstructed, the face was fitted to the cranium in an almost vertical position, much like the flat faces of modern humans. But recent studies of anatomical relationships show that in life the face must have jutted out considerably, creating an ape-like aspect, rather like the faces of *Australopithecus*.[75]

The evolutionist paleoanthropologist J. E. Cronin states the following on the matter:

> ... its relatively robustly constructed face, flattish naso-alveolar clivus, (recalling australopithecine dished faces), low maximum cranial width (on the tem-

porals), strong canine juga and large molars (as indicated by remaining roots) are all relatively primitive traits which ally the specimen with members of the taxon *A. africanus*.[76]

C. Loring Brace from Michigan University came to the same conclusion. As a result of the analyses he conducted on the jaw and tooth structure of skull 1470, he reported that "from the size of the palate and the expansion of the area allotted to molar roots, it would appear that ER 1470 retained a fully *Australopithecus*-sized face and dentition".[77]

Professor Alan Walker, a paleoanthropologist from Johns Hopkins University who has done as much research on KNM-ER 1470 as Leakey, maintains that this creature should not be classified as a member of *Homo*- i.e., as a human species-but rather should be placed in the *Australopithecus* genus.[78]

In summary, classifications like **Homo habilis or *Homo rudolfensis*** which are presented as transitional links between the australopithecines and *Homo erectus* are entirely imaginary. It has been confirmed by many researchers today that these creatures **are members of the *Australopithecus* series**. All of their anatomical features reveal that they are species of ape.

This fact has been further established by two evolutionist anthropologists, Bernard Wood and Mark Collard, whose research was published in 1999 in *Science* magazine. Wood and Collard explained that the *Homo habilis* and *Homo rudolfensis* (Skull 1470) taxa are imaginary, and that the fossils assigned to these categories should be attributed to the genus *Australopithecus*:

> More recently, fossil species have been assigned to *Homo* on the basis of absolute brain size, inferences about language ability and hand function, and retrodictions about their ability to fashion stone tools. With only a few exceptions, the definition and use of the genus within human evolution, and the demarcation of *Homo*, have been treated as if they are unproblematic. But ... recent data, fresh interpretations of the existing evidence, and the limitations of the paleoanthropological record invalidate existing criteria for attributing taxa to *Homo*.
>
> ...in practice fossil hominin species are assigned to *Homo* on the basis of one or more out of four criteria. ... It is now evident, however, that none of these criteria is satisfactory. The Cerebral Rubicon is problematic because absolute cranial capacity is of questionable biological significance. Likewise, there is compelling evidence that language function cannot be reliably inferred from the gross appearance of the brain, and that the language-related parts of the

brain are not as well localized as earlier studies had implied...

...In other words, with the hypodigms of *H. habilis* and *H. rudolfensis* assigned to it, the genus *Homo* is not a good genus. Thus, *H. habilis* and *H. rudolfensis* (or *Homo habilis sensu lato* for those who do not subscribe to the taxonomic subdivision of "early *Homo*") **should be removed** from *Homo*. The obvious taxonomic alternative, which is to transfer one or both of the taxa to one of the existing early hominin genera, is not without problems, but **we recommend that, for the time being, both *H. Habilis* and *H. Rudolfensis* should be transferred to the genus *Australopithecus*.**[79]

The conclusion of Wood and Collard corroborates the conclusion we have maintained here:"Primitive human ancestors" do not exist in history. Creatures that are alleged to be so are actually apes that ought to be assigned to the genus *Australopithecus*. The fossil record shows that there is no evolutionary link between these extinct apes and *Homo*, i.e., human species that suddenly appears in the fossil record.

Homo Erectus and Thereafter: Human Beings

According to the fanciful scheme suggested by evolutionists, the internal evolution of the *Homo* genus is as follows: First *Homo erectus*, then so-called "archaic" *Homo sapiens* and Neanderthal man (*Homo sapiens neanderthalensis*), and finally, Cro-Magnon man (*Homo sapiens sapiens*). However all these classifications are really only variations and unique races in the human family. The difference between them is no greater than the difference between an Inuit and an African or a pygmy and a European.

Let us first examine *Homo erectus*, which is referred to as the most primitive human species. As the name implies, "*Homo erectus*" means "man who walks upright". Evolutionists have had to separate these fossils from earlier ones by adding the qualification of "erectness", because all the available *Homo erectus* fossils are straight to an extent not observed in any of the australopithecines or so-called *Homo habilis* specimens. **There is no difference between the postcranial skeleton of modern man and that of *Homo erectus*.**

The primary reason for evolutionists' defining *Homo erectus* as "primitive", is the cranial capacity of its skull (900-1,100 cc), which is smaller than the average modern man, and its thick eyebrow projections. However, **there are many people living today in the world who have the same cranial capacity as *Homo erectus*** (pygmies, for instance) and other races have protruding eyebrows (Native Australians, for instance).

Homo erectus: An Ancient Human Race

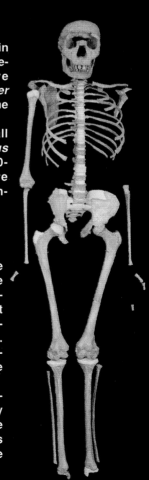

Homo erectus means "upright man". All the fossils included in this species belong to particular human races. Since most of the *Homo erectus* fossils do not have a common characteristic, it is quite hard to define these men according to their skulls. This is the reason why different evolutionist researchers have made various classifications and designations. Above left is seen a skull which was found in Koobi Fora, Africa in 1975 which may generally define *Homo erectus*. Above right is a skull, *Homo ergaster* KNM-ER 3733, which has the obscurities in question.
The cranial capacities of all these diverse *Homo erectus* fossils surge between 900-1100 cc. These figures are within the limits of the contemporary human cranial capacity.

KNM-WT 15000 or Turkana Child skeleton on the right, is probably the oldest and the most complete human fossil ever found. Research made on this fossil which is said to be 1.6 million year old shows that this belongs to a 12 year old child who would become around 1.80 m. tall if he reached adolescence. This fossil which very much resembled to the Neanderthal race, is one of the most remarkable evidence invalidating the story of human's evolution.
The evolutionist Donald Johnson describes this fossil as follows: "He was tall and skinny. His body shape and the proportion of his limbs were the same as the current Equator Africans. The sizes of his limbs totally matched with that of the current white North American adults."

It is a commonly agreed-upon fact that differences in cranial capacity do not necessarily denote differences in intelligence or abilities. Intelligence depends on the internal organisation of the brain, rather than on its volume.[80]

The fossils that have made *Homo erectus* known to the entire world are those of **Peking man** and **Java man** in Asia. However, in time it was realised that these two fossils are not reliable. Peking Man consists of some elements made of plaster whose originals have been lost, and Java Man is "composed" of a skull fragment plus a pelvic bone that was found metres away from it with no indication that these belonged to the same creature. This is why the *Homo erectus* fossils found in Africa have gained such increasing importance. (It should also be noted that some of the fossils said to be *Homo erectus* were included under a second species named "***Homo ergaster***" by some evolutionists. There is disagreement among the experts on this issue. We will treat all these fossils under the classification of *Homo erectus*)

The most famous of the *Homo erectus* specimens found in Africa is the fossil of "*Narikotome Homo erectus*" or the "**Turkana Boy**" which was found near Lake Turkana in Kenya. It is confirmed that the fossil was that of a 12-year-old boy, who would have been 1.83 meters tall in adolescence. The upright skeletal structure of the fossil is no different from that of modern man. The American paleoanthropologist Alan Walker said that he doubted that "the average pathologist could tell the difference between the fossil skeleton and that of a modern human."[81] Concerning the skull, Walker wrote that he laughed when he saw it because "it looked so much like a Neanderthal."[82] As we will see in the next chapter, Neanderthals are a modern human race. Therefore, *Homo erectus* is also a modern human race.

Even the evolutionist Richard Leakey states that the differences between *Homo erectus* and modern man are no more than racial variance:

> One would also see differences in the shape of the skull, in the degree of protrusion of the face, the robustness of the brows and so on. **These differences are probably no more pronounced than we see today between the separate geographical races of modern humans.** Such biological variation arises when populations are geographically separated from each other for significant lengths of time.[83]

Professor William Laughlin from the University of Connecticut made extensive anatomical examinations of Inuits and the people living on the Aleut islands, and noticed that these people were extraordinarily similar to

"Early humans were much smarter than we suspected..."
News published in *New Scientist* on March 14th 1998 tells us that the humans called *Homo Erectus* by evolutionists were practicing seamanship 700 thousand years ago. These humans, who had enough knowledge and technology to build a vessel and possess a culture that made use of sea transport, can hardly be called "primitive".

Homo erectus. The conclusion Laughlin arrived at was that all these distinct races were in fact different races of *Homo sapiens* (modern man).

> When we consider the vast differences that exist between remote groups such as Eskimos and Bushmen, who are known to belong to the single species of *Homo sapiens*, it seems justifiable to conclude that *Sinanthropus* [an *erectus* specimen] belongs within this same diverse species.[84]

It is now a more pronounced fact in the scientific community that *Homo erectus* is a superfluous taxon, and that fossils assigned to the *Homo erectus* class are actually not so different from *Homo sapiens* as to be considered a different species. In *American Scientist,* the discussions over this issue and the result of a conference held on the subject in 2000 were summarised in this way:

> Most of the participants at the Senckenberg conference got drawn into a flaming debate over the taxonomic status of *Homo erectus* started by Milford Wolpoff of the University of Michigan, Alan Thorne of the University of Canberra and their colleagues. They argued forcefully that *Homo erectus* had no validity as a species and should be eliminated altogether. All members of the genus *Homo*, from about 2 million years ago to the present, were one highly variable, widely spread species, *Homo sapiens*, with no natural breaks or subdivisions. The subject of the conference, *Homo erectus* didn't exist.[85]

The conclusion reached by the scientists defending the abovementioned thesis can be summarised as "*Homo erectus* is not a different species from *Homo sapiens*, but rather a race within *Homo sapiens*".

On the other hand, there is a huge gap between *Homo erectus*, a human race, and the apes that preceded *Homo erectus* in the "human evolution" scenario, (*Australopithecus*, *Homo Habilis*, and *Homo rudolfensis*). This means that the first men appeared in the fossil record suddenly and without any prior evolutionary history. There can be no clearer indication of their being created.

Yet, admitting this fact is totally against the dogmatic philosophy and ideology of evolutionists. As a result, they try to portray *Homo erectus*, a truly human race, as a half-ape creature. In their *Homo erectus* reconstructions, they tenaciously draw simian features. On the other hand, with similar drawing methods, they humanise apes like *Australopithecus* or *Homo Habilis*. With this method, they seek to "approximate" apes and human beings and close the gap between these two distinct living classes.

Neanderthals

Neanderthals were human beings who suddenly appeared 100,000 years ago in Europe, and who disappeared, or were assimilated by mixing with other races, quietly but quickly 35,000 years ago. Their only difference from modern man is that their skeletons are more robust and their cranial capacity slightly bigger.

Neanderthals were a human race, a fact which is admitted by almost everybody today. Evolutionists have tried very hard to present them as a "primitive species", yet all the findings indicate that they were no different from a "robust" man walking on the street today. A prominent authority on the subject, Erik Trinkaus, a paleoanthropologist from New Mexico University writes:

FALSE MASKS: Although no different from modern man, Neanderthals are still depicted as ape-like by evolutionists.

> Detailed comparisons of Neanderthal skeletal remains with those of modern humans have shown that **there is nothing in Neanderthal anatomy that conclusively indicates** locomotor, manipulative, intellectual, or linguistic **abilities inferior to those of modern humans.**[86]

Many contemporary researchers define Neanderthal man as a subspecies of modern man and call him "*Homo sapiens neandertalensis*". The findings testify that Neanderthals buried their dead, fashioned musical instruments, and had cultural affinities with the *Homo sapiens sapiens* living during the same period. To put it precisely, Neanderthals are a "robust" human race that simply disappeared in time.

Homo Sapiens Archaic, *Homo Heilderbergensis* and Cro-Magnon Man

Archaic *Homo sapiens* is the last step before contemporary man in the imaginary evolutionary scheme. In fact, evolutionists do not have much to say about these fossils, as there are only very minor differences between them and modern human beings. Some researchers even state that representatives of this race are still living today, and point to native Australians as an example. Like *Homo sapiens* (archaic), native Australians also have thick protruding eyebrows, an inward-inclined mandibular structure, and a slightly smaller cranial capacity.

The group characterised as *Homo heilderbergensis* in evolutionist literature is in fact the same as archaic *Homo sapiens*. The reason why two different terms are used to define the same human racial type is the disagreements among evolutionists. All the fossils included under the *Homo heidelbergensis* classification suggest that people who were anatomically very similar to modern Europeans lived 500,000 and even 740,000 years ago, first in England and then in Spain.

It is estimated that Cro-Magnon man lived 30,000 years ago. He has a dome-shaped cranium and a broad forehead. His cranium of 1,600 cc is above the average for contemporary man. His skull has thick eyebrow projections and a bony protrusion at the back that is characteristic of both Neanderthal man and *Homo erectus*.

Although the Cro-Magnon is considered to be a European race, the structure and volume of Cro-Magnon's cranium look very much like those of some races living in Africa and the tropics today. Relying on this similarity, it is estimated that Cro-Magnon was an archaic African race. Some other paleoanthropological finds have shown that the Cro-Magnon and the Neanderthal races intermixed and laid the foundations for the races of our day.

As a result, none of these human beings were "primitive species".

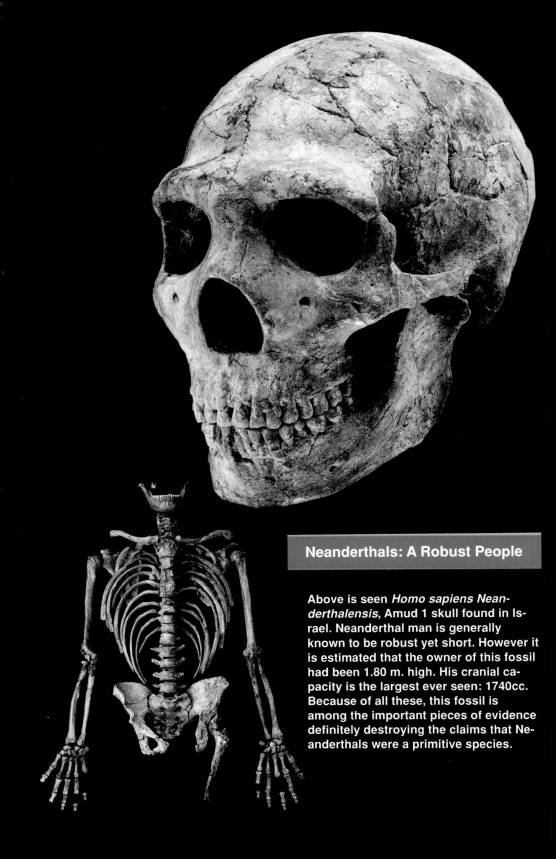

Neanderthals: A Robust People

Above is seen *Homo sapiens Neanderthalensis*, Amud 1 skull found in Israel. Neanderthal man is generally known to be robust yet short. However it is estimated that the owner of this fossil had been 1.80 m. high. His cranial capacity is the largest ever seen: 1740cc. Because of all these, this fossil is among the important pieces of evidence definitely destroying the claims that Neanderthals were a primitive species.

They were different human beings who lived in earlier times and either assimilated and mixed with other races, or became extinct and disappeared from history.

Species Living in the Same Age as Their Ancestors

What we have investigated so far forms a clear picture: The scenario of "human evolution" is a complete fiction. In order for such a family tree to represent the truth, a gradual evolution from ape to man must have taken place and a fossil record of this process should be able to be found. In fact, however, **there is a huge gap between apes and humans**. Skeletal structures, cranial capacities, and such criteria as walking upright or bent sharply forward distinguish humans from apes. (We already mentioned that on the basis of recent research done in 1994 on the inner ear, *Australopithecus* and *Homo habilis* were reclassified as apes, while *Homo erectus* was reclassified as a fully modern human.)

Another significant finding proving that there can be no family-tree relationship among these different species is that species that are presented as ancestors of others in fact lived concurrently. If, as evolutionists claim, *Australopithecus* changed into *Homo habilis*, which, in turn, turned into *Homo erectus*, the periods they lived in should necessarily have followed each other. However, there is no such chronological order to be seen in the fossil record.

According to evolutionist estimates, *Australopithecus* lived from 4 million up until 1 million years ago. The creatures classified as *Homo habilis*, on the other hand, are thought to have lived until 1.7 to 1.9 million years ago. *Homo rudolfensis*, which is said to have been more "advanced" than *Homo habilis*, is known to be as old as from 2.5 to 2.8 million years! That is to say, *Homo rudolfensis* is nearly 1 million years older than *Homo habilis*, of which it is alleged to have been the "ancestor". On the other hand, the age of *Homo erectus* goes as far back as 1.6-1.8 million years ago, which means that *Homo erectus* appeared on the earth in the same time frame as its so-called ancestor, *Homo habilis*.

Alan Walker confirms this fact by stating that "there is evidence from East Africa for late-surviving small ***Australopithecus* individuals that were contemporaneous first with *H. Habilis*, then with *H. erectus*.**"[87] Louis Leakey has found fossils of *Australopithecus*, *Homo habilis* and *Homo*

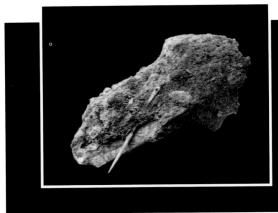

26,000 YEAR OLD NEEDLE: An interesting fossil showing that the Neanderthals had knowledge of clothing: A needle 26,000 years old. (D. Johanson, B. Edgar *From Lucy to Language*, p. 99)

erectus almost next to each other in the Olduvai Gorge region of Tanzania, in the Bed II layer.[88]

There is definitely no such family tree. Stephen Jay Gould, who was a paleontologist from Harvard University, explained this deadlock faced by evolution, although he was an evolutionist himself:

> **What has become of our ladder** if there are three coexisting lineages of hominids (*A. africanus*, the robust australopithecines, and *H. habilis*), none clearly derived from another? Moreover, none of the three display any evolutionary trends during their tenure on earth.[89]

When we move on from *Homo erectus* to *Homo sapiens*, we again see that there is no family tree to talk about. There is evidence showing that *Homo erectus* and archaic *Homo sapiens* continued living up to 27,000 years and even as recently as 10,000 years before our time. In the Kow Swamp in Australia, some 13,000-year-old *Homo erectus* skulls have been found. On the island of Java, *Homo erectus* remains were found that are 27,000 years old.[90]

The Secret History of *Homo Sapiens*

The most interesting and significant fact that nullifies the very basis of the imaginary family tree of evolutionary theory is the **unexpectedly ancient history of modern man**. Paleoanthropological findings reveal that *Homo sapiens* people who looked exactly like us were living as long as 1 million years ago.

It was Louis Leakey, the famous evolutionist paleoanthropologist, who discovered the first findings on this subject. In 1932, in the Kanjera region around Lake Victoria in Kenya, Leakey found several fossils that belonged to the Middle Pleistocene and that were no different from modern man. However, the Middle Pleistocene was a million years ago.⁹¹ Since these discoveries turned the evolutionary family tree upside down, they were dismissed by some evolutionist paleoanthropologists. Yet Leakey always contended that his estimates were correct.

Just when this controversy was about to be forgotten, a fossil unearthed in Spain in 1995 revealed in a very remarkable way that the history of *Homo sapiens* was much older than had been assumed. The fossil in question was uncovered in a cave called Gran Dolina in the **Atapuerca** region of Spain by three Spanish paleoanthropologists from the University of Madrid. The fossil revealed the face of an 11-year-old boy who looked entirely like modern man. Yet, it had been 800,000 years since the child died. *Discover* magazine covered the story in great detail in its December 1997 issue.

This fossil even shook the convictions of Juan Luis Arsuaga Ferreras, who lead the Gran Dolina excavation. Ferreras said:

> We expected something big, something large, something inflated-you know, something primitive. Our expectation of an 800,000-year-old boy was something like Turkana Boy. And what we found was a totally modern face.... To me this is most spectacular-these are the kinds of things that shake you. Finding something totally unexpected like that. Not finding fossils; finding fossils is unexpected too, and it's okay. But the most spectacular thing is finding something you thought belonged to the present, in the past. It's like finding something like-like a tape recorder in Gran Dolina. That would be very surprising. **We don't expect cassettes and tape recorders in the Lower Pleistocene. Finding a modern face 800,000 years ago-it's the same thing.** We were very surprised when we saw it.⁹²

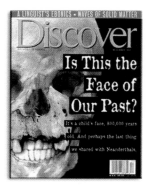

One of the most popular periodicals of the evolutionist literature, *Discover*, put the 800 thousand-year-old human face on its cover with the evolutionists' question "Is this the face of our past?".

The fossil highlighted the fact that the history of *Homo sapiens* had to be extended back to 800,000 years ago. After recovering from the initial shock, the evolutionists who discovered the fossil decided that it belonged to a different species, because according to the evolutionary family tree, *Homo sapiens* did not live 800,000 years ago. Therefore, they made up an imaginary species called "*Homo antecessor*" and included the Atapuerca skull under this classification.

A Hut 1.7 Million Years Old

There have been many findings demonstrating that *Homo sapiens* dates back even earlier than 800,000 years. One of them is a discovery by Louis Leakey in the early 1970s in Olduvai Gorge. Here, in the Bed II layer, Leakey discovered that *Australopithecus, Homo Habilis* and *Homo erectus* species had co-existed at the same time. What is even more interesting was a structure Leakey found in the same layer (Bed II). Here, he found the remains of **a stone hut**. The unusual aspect of the event was that this construction, which is still used in some parts of Africa, could only have been built by *Homo sapiens*! So, according to Leakey's findings, *Australopithecus, Homo habilis, Homo erectus* and modern man must have co-existed approximately 1.7 million years ago.[93] This discovery must surely invalidate the evolutionary theory that claims that modern men evolved from ape-like species such as *Australopithecus*.

Findings of a 1.7 million-year-old hut shocked the scientific community. It looked like the huts used by some Africans today.

Footprints of Modern Man, 3.6 Million Years Old!

Indeed, some other discoveries trace the origins of modern man back to 1.7 million years ago. One of these important finds is the footprints found in Laetoli, Tanzania, by Mary Leakey in 1977. These footprints were found in a layer that was calculated to be 3.6 million years old, and more importantly, they were no different from the footprints that a contemporary man would leave.

The footprints found by Mary Leakey were later examined by a number of famous paleoanthropologists, such as Donald Johanson and Tim White. The results were the same. White wrote:

> Make no mistake about it, ...**They are like modern human footprints.** If one were left in the sand of a California beach today, and a four-year old were asked what it was, he would instantly say that somebody had walked there. He wouldn't be able to tell it from a hundred other prints on the beach, nor would you.[94]

After examining the footprints, Louis Robbins from the University of North California made the following comments:

The arch is raised-the smaller individual had a higher arch than I do-and the big toe is large and aligned with the second toe... The toes grip the ground like human toes. You do not see this in other animal forms.[95]

Examinations of the morphological form of the footprints showed time and again that they had to be accepted as the prints of a human, and moreover, a modern human (*Homo sapiens*). Russell Tuttle, who also examined the footprints wrote:

The Laetoli footprints belonged to modern humans, however they were millions of years old.

> **A small barefoot *Homo sapiens* could have made them**... In all discernible morphological features, the feet of the individuals that made the trails are indistinguishable from those of modern humans.[96]

Impartial examinations of the footprints revealed their real owners. In reality, these footprints consisted of 20 fossilised footprints of a 10-year-old modern human and 27 footprints of an even younger one. They were certainly modern people like us.

This situation put the Laetoli footprints at the centre of discussions for years. Evolutionist paleoanthropologists desperately tried to come up with an explanation, as it was hard for them to accept the fact that a modern man had been walking on the earth 3.6 million years ago. During the 1990s, the following "explanation" started to take shape: The evolutionists decided that these footprints must have been left by an *Australopithecus*, because according to their theory, it was impossible for a *Homo species* to have existed 3.6 years ago. However, Russell H. Tuttle wrote the following in an article in 1990:

> In sum, the 3.5-million-year-old footprint traits at Laetoli site G resemble those of habitually unshod modern humans. None of their features suggest that the Laetoli hominids were less capable bipeds than we are. If the G footprints were not known to be so old, we would readily conclude that there had been made by a member of our genus, *Homo*... In any case, we should shelve the loose assumption that the Laetoli footprints were made by Lucy's kind, *Australopithecus afarensis*.[97]

To put it briefly, these footprints that were supposed to be 3.6 million years old could not have belonged to *Australopithecus*. The only reason

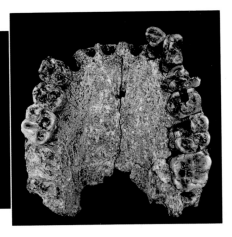

Another example showing the invalidity of the imaginary family tree devised by evolutionists: a modern human (*Homo sapiens*) mandible aged 2.3 million years. This mandible coded A.L. 666-1 was unearthed in Hadar, Ethiopia. Evolutionist publications seek to gloss it over by referring to it as "a very startling discovery"... (*D. Johanson, Blake Edgar, From Lucy to Language, p.169*)

why the footprints were thought to have been left by members of *Australopithecus* was the 3.6-million-year-old volcanic layer in which the footprints were found. The prints were ascribed to *Australopithecus* purely on the assumption that humans could not have lived so long ago.

These interpretations of the Laetoli footprints demonstrate one important fact. Evolutionists support their theory not based on scientific findings, but in spite of them. Here we have a theory that is blindly defended no matter what, with all new findings that cast the theory into doubt being either ignored or distorted to support the theory.

Briefly, the theory of evolution is not science, but a dogma kept alive despite science.

The Bipedalism Impasse of Evolution

Apart from the fossil record that we have dealt with so far, unbridgeable anatomical gaps between men and apes also invalidate the fiction of human evolution. One of these has to do with the manner of walking.

Human beings walk upright on two feet. This is a very special form of locomotion not seen in any other mammalian species. Some other animals do have a limited ability to move when they stand on their two hind feet. Animals like bears and monkeys can move in this way only rarely, such as when they want to reach a source of food, and even then only for a short time. Normally, their skeletons lean forward and they walk on all fours.

Well, then, has bipedalism evolved from the quadrupedal gait of apes, as evolutionists claim?

Of course not. Research has shown that the evolution of bipedalism never occurred, nor is it possible for it to have done so. First of all, bipedalism is not an evolutionary advantage. The way in which monkeys move is much easier, faster, and more efficient than man's bipedal stride. Man can neither move by jumping from tree to tree without descending to the ground, like a chimpanzee, nor run at a speed of 125 km per hour, like a cheetah. On the contrary, since man walks on two feet, he moves much more slowly on the ground. For the same reason, he is one of the most unprotected of all species in nature in terms of movement and defence. According to the logic of evolution, monkeys should not have evolved to adopt a bipedal stride; humans should instead have evolved to become quadrupedal.

Another impasse of the evolutionary claim is that bipedalism does not serve the "gradual development" model of Darwinism. This model, which constitutes the basis of evolution, requires that there should be a "compound" stride between bipedalism and quadrupedalism. However, with the computerised research he conducted in 1996, the English paleoanthropologist Robin Crompton, showed that such a "compound" stride was not possible. Crompton reached the following conclusion: A living being can either walk upright, or on all fours.[98] A type of stride between the two is impossible because it would involve excessive energy consumption. This is why a half-bipedal being cannot exist.

The immense gap between man and ape is not limited solely to bipedalism. Many other issues still remain unexplained, such as brain capacity, the ability to talk, and so on. Elaine Morgan, an evolutionist paleoanthropologist, makes the following confession in relation to this matter:

> Four of the most outstanding mysteries about humans are: 1) why do they walk on two legs? 2) why have they lost their fur? 3) why have they developed such large brains? 4) why did they learn to speak?
> The orthodox answers to these questions are: 1) 'We do not yet know'; 2) 'We do not yet know'; 3) 'We do not yet know'; 4) 'We do not yet know'. The list of questions could be considerably lengthened without affecting the monotony of the answers.[99]

Evolution: An Unscientific Faith

Lord Solly Zuckerman is one of the most famous and respected scientists in the United Kingdom. For years, he studied the fossil record and con-

Recent researches reveal that it is impossible for the bent ape skeleton fit for quadrupedal stride to evolve into upright human skeleton fit for bipedal stride.

ducted many detailed investigations. He was elevated to the peerage for his contributions to science. Zuckerman is an evolutionist. Therefore, his comments on evolution can not be regarded as ignorant or prejudiced. After years of research on the fossils included in the human evolution scenario however, he reached the conclusion that there is no truth to the family tree in that is put forward.

Zuckerman also advanced an interesting concept of the "spectrum of the sciences", ranging from those he considered scientific to those he considered unscientific. According to Zuckerman's spectrum, the most "scientific"-that is, depending on concrete data-fields are chemistry and physics. After them come the biological sciences and then the social sciences. At the far end of the spectrum, which is the part considered to be most "unscientific", are "extra-sensory perception"-concepts such as telepathy and the "sixth sense"-and finally "human evolution". Zuckerman explains his reasoning as follows:

> We then move right off the register of objective truth into those fields of presumed biological science, like extrasensory perception or the **interpretation of man's fossil history, where to the faithful anything is possible** - and where the ardent believer is sometimes able to believe several contradictory things at the same time.[100]

Robert Locke, the editor of *Discovering Archeology*, an important publication on the origins of man, writes in that journal, "The search for human ancestors gives more heat than light", quoting the confession of the famous evolutionist paleoantropologist Tim White:

> We're all frustrated by "all the questions we haven't been able to answer."[101]

Locke's article reviews the impasse of the theory of evolution on the origins of man and the groundlessness of the propaganda spread about this subject:

> Perhaps no area of science is more contentious than the search for human origins. Elite paleontologists disagree over even the most basic outlines of the human family tree. New branches grow amid great fanfare, only to wither and die in the face of new fossil finds.[102]

The same fact was also recently accepted by Henry Gee, the editor of the well-known journal *Nature*. In his book *In Search of Deep Time*, published in 1999, Gee points out that all the evidence for human evolution "between about 10 and 5 million years ago-several thousand generations of living creatures-can be fitted into a small box." He concludes that conventional theories of the origin and development of human beings are "a completely human invention created after the fact, shaped to accord with human prejudices" and adds:

> To take a line of fossils and claim that they represent a lineage is not a scientific hypothesis that can be tested, but an assertion that carries the same validity as bedtime story-amusing, perhaps even instructive, but not scientific.[103]

What, then, is the reason that make so many scientists so tenacious about this dogma? Why have they been trying so hard to keep their theory alive, at the cost of having to admit countless conflicts and discarding the evidence they have found?

The only answer is their being afraid of the fact they will have to face in case of abandoning the theory of evolution. The fact they will have to face when they abandon evolution is that God has created man. However, considering the presuppositions they have and the materialistic philosophy they believe in, creation is an unacceptable concept for evolutionists.

For this reason, they deceive themselves, as well as the world, by using the media with which they co-operate. If they cannot find the necessary fossils, they "fabricate" them either in the form of imaginary pictures or fictitious models and try to give the impression that there indeed exist fossils verifying evolution. A part of mass media who share their materialistic point of view also try to deceive the public and instil the story of evolution in people's subconscious.

No matter how hard they try, the truth is evident: Man has come into existence not through an evolutionary process but by God's creation. Therefore, he is responsible to Him however unwilling he may be to assume this responsibility.

Chapter 10

The Molecular Impasse of Evolution

In previous sections of this book, we have shown how the fossil record invalidates the theory of evolution. In point of fact, there was no need for us to relate any of that, because the theory of evolution collapses long before one gets to any claims about the evidence of fossils. The subject that renders the theory meaningless from the very outset is the question of how life first appeared on earth.

When it addresses this question, evolutionary theory claims that life started with a cell that formed by chance. According to this scenario, four billion years ago various lifeless chemical compounds underwent a reaction in the primordial atmosphere on the earth in which the effects of thunderbolts and atmospheric pressure led to the formation of the first living cell.

The first thing that must be said is that the claim that inanimate materials can come together to form life is an unscientific one that has not been verified by any experiment or observation. Life is only generated from life. Each living cell is formed by the replication of another cell. No one in the world has ever succeeded in forming a living cell by bringing inanimate materials together, not even in the most advanced laboratories.

The theory of evolution claims that a living cell-which cannot be produced even when all the power of the human intellect, knowledge and technology are brought to bear-nevertheless managed to form by chance under primordial conditions of the earth. In the following pages, we will examine why this claim is contrary to the most basic principles of science and reason.

The Tale of the "Cell Produced by Chance"

If one believes that a living cell can come into existence by coincidence, then there is nothing to prevent one from believing a similar story that we will relate below. It is the story of a town:

One day, a lump of clay, pressed between the rocks in a barren land, becomes wet after it rains. The wet clay dries and hardens when the sun rises, and takes on a stiff, resistant form. Afterwards, these rocks, which also served as a mould, are somehow smashed into pieces, and then a neat, well shaped, and strong brick appears. This brick waits under the same natural conditions for years for a similar brick to be formed. This goes on until hundreds and thousands of the same bricks have been formed in the same place. However, by chance, none of the bricks that were previously formed are damaged. Although exposed to storm, rain, wind, scorching sun, and freezing cold for thousands of years, the bricks do not crack, break up, or get dragged away, but wait there in the same place with the same determination for other bricks to form.

When the number of bricks is adequate, they erect a building by being arranged sideways and on top of each other, having been randomly dragged along by the effects of natural conditions such as winds, storms, or tornadoes. Meanwhile, materials such as cement or soil mixtures form under "natural conditions", with perfect timing, and creep between the bricks to clamp them to each other. While all this is happening, iron ore under the ground is shaped under "natural conditions" and lays the foundations of a building that is to be formed with these bricks. At the end of this process, a complete building rises with all its materials, carpentry, and installations intact.

Of course, a building does not only consist of foundations, bricks, and cement. How, then, are the other missing materials to be obtained? The answer is simple: all kinds of materials that are needed for the construction of the building exist in the earth on which it is erected. Silicon for the glass, copper for the electric cables, iron for the columns, beams, water pipes, etc. all exist under the ground in abundant quantities. It takes only the skill of "natural conditions" to shape and place these materials inside the building. All the installations, carpentry, and accessories are placed among the bricks with the help of the blowing wind, rain, and earthquakes. Everything has gone so well that the bricks are arranged so as to leave the necessary window spaces as if they knew that something called glass would be formed later on by natural conditions. Moreover, they have not forgotten to leave some space to allow the installation of water, electricity and heating systems, which are also later to be formed by coincidence. Everything has gone so well that "coincidences" and "natural conditions" produce a perfect design.

FOCUS: Confessions from Evolutionists

The theory of evolution faces no greater crisis than on the point of explaining the emergence of life. The reason is that organic molecules are so complex that their formation cannot possibly be explained as being coincidental and it is manifestly impossible for an organic cell to have been formed by chance.

Evolutionists confronted the question of the origin of life in the second quarter of the 20th century. One of the leading authorities of the theory of molecular evolution, the Russian evolutionist Alexander I. Oparin, said this in his book *The Origin of Life*, which was published in 1936:

> Unfortunately, the origin of the cell remains a question which is actually the darkest point of the complete evolution theory.[1]

Alexander Oparin:

Since Oparin, evolutionists have performed countless experiments, conducted research, and made observations to prove that a cell could have been formed by chance. However, every such attempt only made clearer the complex design of the cell and thus refuted the evolutionists' hypotheses even more. Professor Klaus Dose, the president of the Institute of Biochemistry at the University of Johannes Gutenberg, states:

> More than 30 years of experimentation on the origin of life in the fields of chemical and molecular evolution have led to a better perception of the immensity of the problem of the origin of life on Earth rather than to its solution. At present all discussions on principal theories and experiments in the field either end in stalemate or in a confession of ignorance.[2]

Jeffrey Bad

The following statement by the geochemist Jeffrey Bada from San Diego Scripps Institute makes clear the helplessness of evolutionists concerning this impasse:

> Today as we leave the twentieth century, we still face the biggest unsolved problem that we had when we entered the twentieth century: How did life originate on Earth?[3]

1- Alexander I. Oparin, *Origin of Life*, (1936) NewYork: Dover Publications, 1953 (Reprint), p.196.
2- Klaus Dose, "The Origin of Life: More Questions Than Answers", *Interdisciplinary Science Reviews, Vol* 13, No. 4, 1988, p. 348
3- Jeffrey Bada, *Earth,* February 1998, p. 40

If you have managed to sustain your belief in this story so far, then you should have no trouble surmising how the town's other buildings, plants, highways, sidewalks, substructures, communications, and transportation systems came about. If you possess technical knowledge and are fairly conversant with the subject, you can even write an extremely "scientific" book of a few volumes stating your theories about "the evolutionary process of a sewage system and its uniformity with the present structures". You may well be honoured with academic awards for your clever studies, and may consider yourself a genius, shedding light on the nature of humanity.

The theory of evolution, which claims that life came into existence by chance, is no less absurd than our story, for, with all its operational systems, and systems of communication, transportation and management, a cell is no less complex than a city.

The Miracle in the Cell and the End of Evolution

The complex structure of the living cell was unknown in Darwin's day and at the time, ascribing life to "coincidences and natural conditions" was thought by evolutionists to be convincing enough.

The technology of the 20th century has delved into the tiniest particles of life and has revealed that the cell is the most complex system mankind has ever confronted. Today we know that the cell contains power stations producing the energy to be used by the cell, factories manufacturing the enzymes and hormones essential for life, a databank where all the necessary information about all products to be produced is recorded, complex transportation systems and pipelines for carrying raw materials and products from one place to another, advanced laboratories and refineries for breaking down external raw materials into their useable parts, and specialised cell membrane proteins to control the incoming and outgoing materials. And these constitute only a small part of this incredibly complex system.

W. H. Thorpe, an evolutionist scientist, acknowledges that **"The most elementary type of cell constitutes a 'mechanism' unimaginably more complex than any machine yet thought up, let alone constructed, by man."**[104]

A cell is so complex that even the high level of technology attained

The Complexity of the Cell

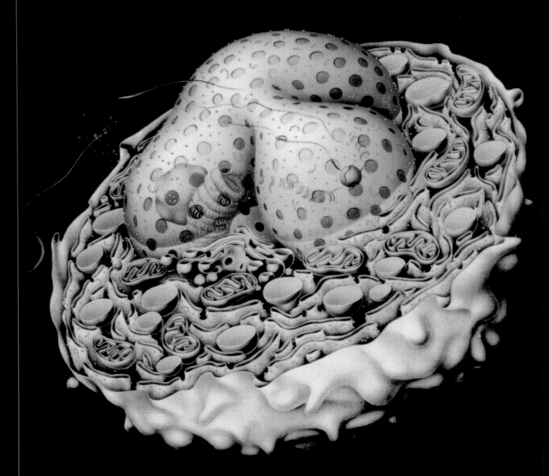

The cell is the most complex and most elegantly designed system man has ever witnessed. Professor of biology Michael Denton, in his book entitled *Evolution: A Theory in Crisis*, explains this complexity with an example:

"To grasp the reality of life as it has been revealed by molecular biology, we must magnify a cell a thousand million times until it is twenty kilometers in diameter and resembles a giant airship large enough to cover a great city like London or New York. What we would then see would be an object of unparalelled complexity and adaptive design. On the surface of the cell we would see millions of openings, like port holes of a vast space ship, opening and closing to allow a continual stream of materials to flow in and out. If we were to enter one of these openings we would find ourselves in a world of supreme technology and bewildering complexity... (a complexity) beyond our own creative capacities, a reality which is the very antithesis of chance, which excels in every sense anything produced by the intelligence of man..."

today cannot produce one. No effort to create an artificial cell has ever met with success. Indeed, all attempts to do so have been abandoned.

The theory of evolution claims that this system-which mankind, with all the intelligence, knowledge and technology at its disposal, cannot succeed in reproducing-came into existence "by chance" under the conditions of the primordial earth. To give another example, the probability of forming of a cell by chance is about the same as that of producing a perfect copy of a book following an explosion in a printing-house.

The English mathematician and astronomer Sir Fred Hoyle made a similar comparison in an interview published in *Nature* magazine on November 12, 1981. Although an evolutionist himself, Hoyle stated that the chance that higher life forms might have emerged in this way is comparable to the chance that **a tornado sweeping through a junk-yard might assemble a Boeing 747 from the materials therein**.[105] This means that it is not possible for the cell to have come into being by coincidence, and therefore it must definitely have been "created".

One of the basic reasons why the theory of evolution cannot explain how the cell came into existence is the "irreducible complexity" in it. A living cell maintains itself with the harmonious co-operation of many organelles. If only one of these organelles fails to function, the cell cannot remain alive. The cell does not have the chance to wait for unconscious mechanisms like natural selection or mutation to permit it to develop. Thus, the first cell on earth was necessarily a complete cell possessing all the required organelles and functions, and this definitely means that this cell had to have been created.

Proteins Challenge Chance

So much for the cell, but evolution fails even to account for the building-blocks of a cell. The formation, under natural conditions, of just one single protein out of the thousands of complex protein molecules making up the cell is impossible.

Proteins are giant molecules consisting of smaller units called "amino acids" that are arranged in a particular sequence in certain quantities and structures. These units constitute the building blocks of a living protein. The simplest protein is composed of 50 amino acids, but there are some that contain thousands.

The crucial point is this. The absence, addition, or replacement of a single amino acid in the structure of a protein causes the protein to become a useless molecular heap. Every amino acid has to be in the right place and in the right order. The theory of evolution, which claims that life emerged as a result of chance, is quite helpless in the face of this order, since it is too wondrous to be explained by coincidence. (Furthermore the theory cannot even substantiate the claim of the accidental formation of amino acids, as will be discussed later.)

The fact that it is quite impossible for the functional structure of proteins to come about by chance can easily be observed even by simple probability calculations that anybody can understand.

For instance, an average-sized protein molecule composed of 288 amino acids, and contains twelve different types of amino acids can be arranged in 10^{300} different ways. (This is an astronomically huge number, consisting of 1 followed by 300 zeros.) Of all these possible sequences, only one forms the desired protein molecule. The rest of them are amino-acid chains that are either totally useless or else potentially harmful to living things.

In other words, the probability of the formation of only one protein molecule is "1 in 10^{300}". The probability of this "1" to occur is practically nil. (In practice, probabilities smaller than 1 over 10^{50} are thought of as "zero probability").

Furthermore, a protein molecule of 288 amino acids is a rather modest one compared with some giant protein molecules consisting of thousands of amino acids. When we apply similar probability calculations to these giant protein molecules, we see that even the word "impossible" is insufficient to describe the true situation.

When we proceed one step further in the evolutionary scheme of life, we observe that one single protein means nothing by itself. One of the smallest bacteria ever discovered, *Mycoplasma hominis* H39, contains 600 "types" of proteins. In this case, we would have to repeat the probability calculations we have made above for one protein for each of these 600 different types of proteins. The result beggars even the concept of impossibility.

Some people reading these lines who have so far accepted the theory of evolution as a scientific explanation may suspect that these numbers are

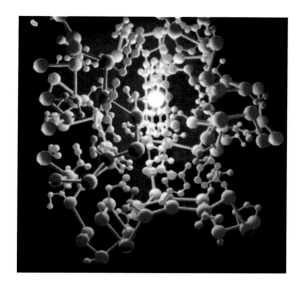

Proteins are the most vital elements for living things. They not only combine to make up living cells, but also play key roles in the body chemistry. From protein synthesis to hormonal communications, it is possible to see proteins in action.

exaggerated and do not reflect the true facts. That is not the case: these are definite and concrete facts. No evolutionist can object to these numbers. They accept that the probability of the coincidental formation of a single protein is "as unlikely as the possibility of a monkey writing the history of humanity on a typewriter without making any mistakes".[106] However, instead of accepting the other explanation, which is creation, they go on defending this impossibility.

This situation is in fact acknowledged by many evolutionists. For example, Harold F. Blum, a prominent evolutionist scientist, states that **"The spontaneous formation of a polypeptide of the size of the *smallest known proteins* seems beyond all probability."** [107]

Evolutionists claim that molecular evolution took place over a very long period of time and that this made the impossible possible. Nevertheless, no matter how long the given period may be, it is not possible for amino acids to form proteins by chance. William Stokes, an American geologist, admits this fact in his book *Essentials of Earth History*, writing that the probability is so small **"that it would not occur during billions of years on billions of planets, each covered by a blanket of concentrated watery solution of the necessary amino acids."** [108]

So what does all this mean? Perry Reeves, a professor of chemistry, answers the question:

When one examines the vast number of possible structures that could result from a simple random combination of amino acids in an evaporating primordial pond, it is mind-boggling to believe that life could have originated in this way. **It is more plausible that a Great Builder with a master plan would be required for such a task.**[109]

If the coincidental formation of even one of these proteins is impossible, it is billions of times "more impossible" for some one million of those proteins to come together properly by chance and make up a complete cell. What is more, by no means does a cell consist of a mere heap of proteins. In addition to the proteins, a cell also includes nucleic acids, carbohydrates, lipids, vitamins, and many other chemicals such as electrolytes arranged in a specific proportion, equilibrium, and design in terms of both structure and function. Each of these elements functions as a building block or co-molecule in various organelles.

Robert Shapiro, a professor of chemistry at New York University and a DNA expert, calculated the probability of the coincidental formation of the 2000 types of proteins found in a single bacterium (There are 200,000 different types of proteins in a human cell). The number that was found was 1 over 10^{40000}.[110] (This is an incredible number obtained by putting 40,000 zeros after the 1)

A professor of applied mathematics and astronomy from University College Cardiff, Wales, Chandra Wickramasinghe, comments:

> **The likelihood of the spontaneous formation of life from inanimate matter is one to a number with 40,000 noughts after it... It is big enough to bury Darwin and the whole theory of evolution.** There was no primeval soup, neither on this planet nor on any other, and if the **beginnings of life** were not random, they must therefore **have been the product of purposeful intelligence.**[111]

Sir Fred Hoyle comments on these implausible numbers:

> Indeed, such a theory (that life was assembled by an intelligence) is so obvious that one wonders why it is not widely accepted as being self-evident. The reasons are psychological rather than scientific.[112]

The reason Hoyle used the term "psychological" is the self-conditioning of evolutionists not to accept that life could have been created. The rejection of God's existence is their main goal. For this reason alone, they go on defending irrational theories which they at the same time acknowledge to be impossible.

Left-handed Proteins

Let us now examine in detail why the evolutionist scenario regarding the formation of proteins is impossible.

Even the correct sequence of the right amino acids is still not enough for the formation of a functional protein molecule. In addition to these requirements, each of the 20 different types of amino acids present in the composition of proteins must be left-handed. There are two different types of amino acids-as of all organic molecules-called "left-handed" and "right-handed". The difference between them is the mirror-symmetry between their three dimensional structures, which is similar to that of a person's right and left hands.

Amino acids of either of these two types can easily bond with one another. But one astonishing fact that has been revealed by research is that all the proteins in plants and animals on this planet, from the simplest organism to the most complex, are made up of left-handed amino acids. If even a single right-handed amino acid gets attached to the structure of a protein, the protein is rendered useless. In a series of experiments, surprisingly, bacteria that were exposed to right-handed amino acids immediately destroyed them. In some cases, they produced usable left-handed amino acids from the fractured components.

Let us for an instant suppose that life came about by chance as evolutionists claim it did. In this case, the right- and left-handed amino acids that were generated by chance should be present in roughly equal proportions in nature. Therefore, all living things should have both right- and left-handed amino acids in their constitution, because chemically it is possible for amino acids of both types to combine with each other. However, as we know, in the real world the proteins existing in all living organisms are made up only of left-handed amino acids.

The question of how proteins can pick out only the left-handed ones from among all amino acids, and how not even a single right-handed amino acid gets involved in the life process, is a problem that still baffles evolutionists. Such a specific and conscious selection constitutes one of the greatest impasses facing the theory of evolution.

Moreover, this characteristic of proteins makes the problem facing evolutionists with respect to "coincidence" even worse. In order for a "meaningful" protein to be generated, it is not enough for the amino acids to be present in a particular number and sequence, and to be combined together in the right three-dimensional design. Additionally, all these amino acids have to be left-handed: not even one of them can be right-handed. Yet there is no natural selection mechanism which can identify that a right-handed amino acid has been added to the sequence and recognise that it must therefore be removed from the chain. This situation once more eliminates for good the possibility of coincidence and chance.

The *Brittanica Science Encyclopaedia,* which is an outspoken defender of evolution, states that the amino acids of all the living organisms on earth, and the building blocks of complex polymers such as proteins, have the same left-handed asymmetry. It adds that this is tantamount to tossing a coin a million times and always getting heads. The same encyclopaedia states that it is impossible to understand why molecules become left-handed or right-handed, and that this choice is fascinatingly related to the origin of life on earth.[113]

If a coin always turns up heads when tossed a million times, is it more logical to attribute that to chance, or else to accept that there is conscious intervention going on? The answer should be obvious. However, obvious though it may be, evolutionists still take refuge in coincidence, simply because they do not want to accept the existence of "conscious intervention".

A situation similar to the left-handedness of amino acids also exists with respect to nucleotides, the smallest units of the nucleic acids, DNA and RNA. In contrast to proteins, in which only left-handed amino acids are chosen, in the case of the nucleic acids, the preferred forms of their nucleotide components are always right-handed. This is another fact that can never be explained by coincidence.

In conclusion, it is proven beyond a shadow of doubt by the probabilities we have examined that the origin of life cannot be explained by chance. If we attempt to calculate the probability of an average-sized protein consisting of 400 amino acids being selected only from left-handed amino acids, we come up with a probability of 1 in 2^{400}, or 10^{120}. Just for a comparison, let us remember that the number of electrons in the universe is estimated at 10^{79}, which although vast, is a much smaller number. The probability of these amino acids forming the required sequence and functional form would generate much larger numbers. If we add these probabilities to each other, and if we go on to work out the probabilities of even higher numbers and types of proteins, the calculations become inconceivable.

Correct Bond is Vital

The difficulties the theory of evolution is unable to overcome with regard to the development of a single protein are not limited to those we have recounted so far. It is not enough for amino acids to be arranged in the correct numbers, sequences, and required three-dimensional structures. The formation of a protein also requires that amino acid molecules with more than one arm be linked to each other only in certain ways. Such a bond is called a "peptide bond". Amino acids can make different bonds with each other; but proteins are made up of those-and only those-amino acids which are joined by "peptide" bonds.

A comparison will clarify this point. Suppose that all the parts of a car were complete and correctly assembled, with the sole exception that one of the wheels was fastened in place not with the usual nuts and bolts, but with a piece of wire, in such a way that its hub faced the ground. It would be impossible for such a car to move even the shortest distance, no matter how complex its technology or how powerful its engine. At first glance, everything would seem to be in the right place, but the faulty attachment of even one wheel would make the entire car useless. In the same way, in a protein molecule the joining of even one amino acid to another with a bond other than a peptide bond would make the entire molecule useless.

Research has shown that amino acids combining at random combine with a peptide bond only 50% of the time, and that the rest of the time different bonds that are not present in proteins emerge. To function properly, each amino acid making up a protein must be joined to others only with a peptide bond, in the same way that it likewise must be chosen only from among left-handed forms.

This probability of this happening is the same as the probability of each protein's being left-handed. That is, when we consider a protein made up of 400 amino acids, the probability of all amino acids combining among themselves with only peptide bonds is 1 in 2^{399}.

Zero Probability

As can be seen below, the probability of formation of a protein molecule made up of 500 amino acids is "1" over a number formed by placing 950 zeros next to 1, which is a number incomprehensible for the human mind. This is a probability only on paper. Practically speaking, there is zero chance of its actually happening. As we saw earlier, in mathematics, a probability smaller than 1 in 10^{50} is statistically considered to have a "0" probability of occurring.

A probability of "1 over 10^{950}" is far beyond the limits of this definition.

While the improbability of the formation of a protein molecule made up of 500 amino acids reaches such an extent, we can further proceed to push the limits of the mind with higher levels of improbability. In the "haemoglobin" molecule, which is a vital protein, there are 574 amino acids, which is more than the amino acids making up the protein mentioned above. Now consider this: in only one out of the billions of red blood cells in your body, there are "280,000,000" (280 million) haemoglobin molecules.

The supposed age of the earth is not sufficient to allow the formation of even a single protein by a "trial and error" method, let alone that of a red blood cell. Even if we suppose that amino acids have combined and decomposed by a "trial and error" method without losing any time since the formation of the earth, in order to form a single protein molecule, the time that would be required for something with a probability of 10^{950} to happen would still hugely exceed the estimated age of the earth.

The conclusion to be drawn from all this is that evolution falls into a terrible abyss of improbability even when it comes to the formation of a single protein.

Is There a Trial and Error Mechanism in Nature?

Finally, we may conclude with a very important point in relation to the basic logic of probability calculations, of which we have already seen some examples. We indicated that the probability calculations made above reach astronomical levels, and that these astronomical odds have no chance of actually happening. However, there is a much more important and damaging fact facing evolutionists here. This is that under natural conditions, no period of trial and error can even start, despite the astronomical odds, because there is no trial-and-error mechanism in nature from which proteins could emerge.

The calculations we give on page across to demonstrate the probability of the formation of a protein molecule with 500 amino acids are valid only for an ideal trial-and-error environment, which does not actually exist in real life.

The Probability of a Protein Being Formed by Chance is Zero

There are 3 basic conditions for the formation of a useful protein:

First condition: that all the amino acids in the protein chain are of the right type and in the right sequence

Second condition: that all the amino acids in the chain are left-handed

Third condition: that all of these amino acids are united between them by forming a chemical bond called "peptide bond".

In order for a protein to be formed by chance, all three basic conditions must exist simultaneously. The probability of the formation of a protein by chance is equal to the multiplication of the probabilities of the realisation of each of these conditions.

For instance, for an average molecule comprising of 500 amino acids:

1. The probability of the amino acids being in the right sequence:

There are 20 types of amino acids used in the composition of proteins. According to this:

- The probability of each amino acid being chosen correctly among these 20 types $= 1/20$
- The probability of all of those 500 amino acids being chosen correctly $= 1/20^{500} = 1/10^{650}$

$= 1$ chance in 10^{650}

2. The probability of the amino acids being left-handed:

- The probability of only one amino acid being left-handed $= 1/2$
- The probability of all of those 500 amino acids being left-handed at the same time $= 1/2^{500} = 1/10^{150}$

$= 1$ chance in 10^{150}

3. The probability of the amino acids being combined with a "peptide bond":

Amino acids can combine with each other with different kinds of chemical bonds. In order for a useful protein to be formed, all the amino acids in the chain must have been combined with a special chemical bond called a "peptide bond". It is calculated that the probability of the amino acids being combined not with another chemical bond but by a peptide bond is 50%. In relation to this:

- The probability of two amino acids being combined with a "peptide bond" $= 1/2$
- The probability of 500 amino acids all combining with peptide bonds $= 1/2^{499} = 1/10^{150}$

$= 1$ chance in 10^{150}

TOTAL PROBABILITY $= 1/10^{650} \times 1/10^{150} \times 1/10^{150} = 1/10^{950}$

$= 1$ chance in 10^{950}

> The probability of an average protein molecule made up of 500 amino acids being arranged in the correct quantity and sequence in addition to the probability of all of the amino acids it contains being only left-handed and being combined with only peptide bonds is "1" over 10^{950}. We can write this number which is formed by putting 950 zeros next to 1 as follows:
>
> $$10^{950} =$$
>
> 100,000

That is, the probability of obtaining a useful protein is "1" in 10^{950} only if we suppose that there exists an imaginary mechanism in which an invisible hand joins 500 amino acids at random and then, seeing that this is not the right combination, disentangles them one by one, and arranges them again in a different order, and so on. In each trial, the amino acids would have to be separated one by one, and be arranged in a new order. The synthesis should be stopped after the 500th amino acid has been added, and it must be ensured that not even one extra amino acid is involved. The trial should then be stopped to see whether or not a functional protein has yet been formed, and, in the event of failure, everything should be split up again and then tested for another sequence. Additionally, in each trial, not even one extraneous substance should be allowed to become involved. It is also imperative that the chain formed during the trial should not be separated and destroyed before reaching the 499th link. These conditions mean that the probabilities we have mentioned above can only operate in a controlled environment where there is a conscious mechanism directing the beginning, the end, and each intermediate stage of the process, and where only "the correct selection of the amino acids" is left to chance. It is clearly impossible for such an environment to exist under natural conditions. Therefore the formation of a protein in the natural environment is logically and technically impossible. In fact, to talk of the probabilities of such an event is quite unscientific.

Since some people are unable to take a broad view of these matters, but approach them from a superficial viewpoint and assume protein formation to be a simple chemical reaction, they may make unrealistic deductions such as "amino acids combine by way of reaction and then form proteins". However, accidental chemical reactions taking place in an inanimate structure can only lead to simple and primitive changes. The number of these is predetermined and limited. For a somewhat more complex chemical material, huge factories, chemical plants, and laboratories have to be involved. Medicines and many other chemical materials that we use in our daily life are made in just this way. Proteins have much more complex structures than these chemicals produced by industry. Therefore, it is impossible for proteins, each of which is a wonder of design and engineering, in which every part takes its place in a fixed order, to originate as a result of haphazard chemical reactions.

Let us for a minute put aside all the impossibilities we have described so far, and suppose that a useful protein molecule still evolved spontaneously "by accident". Even so, evolution again has no answers, because in order for this protein to survive, it would need to be isolated from its natural habitat and be protected under very special conditions. Otherwise, it would either disintegrate from exposure to natural conditions on earth, or else join with other acids, amino acids, or chemical compounds, thereby losing its particular properties and turning into a totally different and useless substance.

The Evolutionary Fuss About the Origin of Life

The question of "how living things first appeared" is such a critical impasse for evolutionists that they usually try not even to touch upon this subject. They try to pass over this question by saying "the first creatures came into existence as a result of some random events in water". They are at a road-block that they can by no means get around. In spite of the paleontological evolution arguments, in this subject they have no fossils available to distort and misinterpret as they wish to support their assertions. Therefore, the theory of evolution is definitely refuted from the very beginning.

Above all, there is one important point to take into consideration: **If any one step in the evolutionary process is proven to be impossible, this jis sufficient to prove that the whole theory is totally false and invalid.** For instance, by proving that the haphazard formation of proteins is impossible, all other claims regarding the subsequent steps of evolution are also refuted. After this, it becomes meaningless to take some human and ape skulls and engage in speculation about them.

How living organisms came into existence out of nonliving matter was an issue that evolutionists did not even want to mention for a long time. However, this question, which had constantly been avoided, eventually had to be addressed, and attempts were made to settle it with a series of experiments in the second quarter of the 20th century.

The main question was: How could the first living cell have appeared in the primordial atmosphere on the earth? In other words, what kind of explanation could evolutionists offer?

The answers to the questions were sought through experiments. Evolutionist scientists and researchers carried out laboratory experiments directed at answering these questions but these did not create much interest. The most generally respected study on the origin of life is the **Miller experiment** conducted by the American researcher Stanley Miller in 1953. (The experiment is also known as "Urey-Miller experiment" because of the contribution of Miller's instructor at the University of Chicago, Harold Urey.)

This experiment is the only "evidence" evolutionists have with which to allegedly prove the "molecular evolution thesis"; they advance it as the first stage of the supposed evolutionary process leading to life. Although nearly half a century has passed, and great technological advances have been made, nobody has made any further progress. In spite of this, Miller's experiment is still taught in textbooks as the evolutionary explanation of the earliest generation of living things. Aware of the fact that such studies do not support, but rather actually refute, their thesis, evolutionist researchers deliberately avoid embarking on such experiments.

Miller's Experiment

Stanley Miller's aim was to demonstrate by means of an experiment that amino acids, the building blocks of proteins, could have come into existence "by chance" on the lifeless earth billions of years ago.

In his experiment, Miller used a gas mixture that he assumed to have existed on the primordial earth (but which later proved unrealistic) composed of ammonia, methane, hydrogen, and water vapour. Since these gasses would not react with each other under natural conditions, he added energy to the mixture to start a reaction among them. Supposing that this energy could have come from lightning in the primordial atmosphere, he used an electric current for this purpose.

Miller heated this gas mixture at 100^0C for a week and added the electrical current. At the end of the week, Miller analysed the chemicals which had formed at the bottom of the jar, and observed that three out of the 20 amino acids, which constitute the basic elements of proteins had been synthesised.

This experiment aroused great excitement among evolutionists, and was promoted as an outstanding success. Moreover, in a state of intoxicated euphoria, various publications carried headlines such as "Miller creates life". However, what Miller had managed to synthesise was only a few "inanimate" molecules.

Encouraged by this experiment, evolutionists immediately produced new scenarios. Stages following the development of amino acids were hurriedly hypothesised. Supposedly, amino acids had later united in the correct sequences by accident to form proteins. Some of these proteins which emerged by chance formed themselves into cell membrane-like structures which "somehow" came into existence and formed a primitive cell. The cells then supposedly came together over time to form multicellular living organisms. However, Miller's experiment was nothing but make-believe and has since proven to be false in many aspects.

Miller's Experiment was Nothing but Make-believe

Miller's experiment sought to prove that amino acids could form on their own in primordial earth-like conditions, but it contains inconsistencies in a number of areas:

1. By using a mechanism called a "cold trap", Miller isolated the amino acids from the environment as soon as they were formed. Had he not done so, the conditions in the environment in which the amino acids were formed would immediately have destroyed these molecules.

Doubtless, this kind of a conscious mechanism of isolation did not exist on the primordial earth. Without such a mechanism, even if one amino acid were obtained, it would immediately have been destroyed. The chemist Richard Bliss expresses this contradiction by observing that "Actually, without this trap, the chemical products would have been destroyed by the energy source."[114]

And, sure enough, in his previous experiments, Miller had been unable to make even one single amino acid using the same materials without the cold trap mechanism.

2. The primordial atmospheric environment that Miller attempted to simulate in his experiment was not realistic. In the 1980s, scientists agreed that nitrogen and carbon dioxide should have been used in this artificial environment instead of methane and ammonia. After a long period of silence, Miller himself also confessed that the atmospheric environment he used in his experiment was not realistic.[115]

So why did Miller insist on these gasses? The answer is simple: without ammonia, it was impossible to synthesise any amino acid. Kevin Mc Kean talks about this in an article published in *Discover* magazine:

> Miller and Urey imitated the ancient atmosphere on the Earth with a mixture of methane and ammonia. According to them, the Earth was a true homogeneous mixture of metal, rock and ice. However in the latest studies, it has been understood that the Earth was very hot at those times, and that it was composed of melted nickel and iron. Therefore, the chemical atmosphere of that time should have been formed mostly of nitrogen (N_2), carbon dioxide (CO_2) and water vapour (H_2O). However these are not as appropriate as methane and ammonia for the production of organic molecules.[116]

The American scientists J.P. Ferris and C.T. Chen repeated Miller's experiment with an atmospheric environment that contained carbon dioxide, hydrogen, nitrogen, and water vapour, and were unable to obtain even a single amino acid molecule.[117]

3. Another important point that invalidates Miller's experiment is that **there was enough oxygen to destroy all the amino acids in the atmosphere at the time when they were thought to have been formed.** This fact,

Latest Evolutionist Sources Dispute Miller's Experiment

FOCUS

Today, Miller's experiment is totally disregarded even by evolutionist scientists. In the February 1998 issue of the famous evolutionist science journal *Earth*, the following statements appear in an article titled "Life's Crucible":

Geologist now think that the primordial atmosphere consisted mainly of carbon dioxide and nitrogen, gases that are less reactive than those used in the 1953 experiment. And even if Miller's atmosphere could have existed, how do you get simple molecules such as amino acids to go through the necessary chemical changes that will convert them into more complicated compounds, or polymers, such as proteins? Miller himself throws up his hands at that part of the puzzle. "It's a problem," he sighs with exasperation. "How do you make polymers? That's not so easy."[1]

As seen, today even Miller himself has accepted that his experiment does not lead to an explanation of the origin of life. The fact that evolutionist scientists embraced this experiment so fervently only indicates the difficulties facing evolution, and the desperation of its advocates.

In the March 1998 issue of *National Geographic*, in an article titled "The Emergence of Life on Earth", the following comments appear:

Many scientists now suspect that the early atmosphere was different from what Miller first supposed. They think it consisted of carbon dioxide and nitrogen rather than hydrogen, methane, and ammonia.

That's bad news for chemists. When they try sparking carbon dioxide and nitrogen, they get a paltry amount of organic molecules - the equivalent of dissolving a drop of food colouring in a swimming pool of water. Scientists find it hard to imagine life emerging from such a diluted soup.[2]

In brief, neither Miller's experiment, nor any other similar one that has been attempted, can answer the question of how life emerged on earth. All of the research that has been done shows that it is impossible for life to emerge by chance, and thus confirms that life is created.

1- Earth, "Life's Crucible", February 1998, p.34
2- National Geographic, "The Rise of Life on Earth", March 1998, p.68

overlooked by Miller, is revealed by the traces of oxidised iron and uranium found in rocks that are estimated to be 3.5 billion years old.[118]

There are other findings showing that the amount of oxygen in the atmosphere at that time was much higher than originally claimed by evolutionists. Studies also show that at that time, the amount of ultraviolet radiation to which the earth was then exposed was 10,000 times more than evolutionists' estimates. This intense radiation would unavoidably have freed oxygen by decomposing the water vapour and carbon dioxide in the atmosphere.

This situation completely negates Miller's experiment, in which oxygen was completely neglected. If oxygen had been used in the experiment, methane would have decomposed into carbon dioxide and water, and ammonia into nitrogen and water. On the other hand, in an environment where there was no oxygen, there would be no ozone layer either; therefore, the amino acids would have immediately been destroyed, since they would have been exposed to the most intense ultraviolet rays without the protection of the ozone layer. In other words, with or without oxygen in the primordial world, the result would have been a deadly environment for the amino acids.

4. At the end of Miller's experiment, many organic acids had been formed with characteristics detrimental to the structure and function of living things. If the amino acids had not been isolated, and had been left in the same environment with these chemicals, their destruction or transformation into different compounds through chemical reactions would have been unavoidable.

Moreover, a large number of right-handed amino acids were formed at the end of the experiment.[119] The existence of these amino acids refuted the theory even within its own terms because right-handed amino acids cannot function in the composition of living organisms. To conclude, the circumstances in which amino acids were formed in Miller's experiment were not suitable for life. In truth, this medium took the form of an acidic mixture destroying and oxidising the useful molecules obtained.

All these facts point to one firm truth: **Miller's experiment cannot claim to have proved that living things formed by chance under primordial earth-like conditions.** The whole experiment is nothing more than a deliberate and controlled laboratory experiment to synthesise amino acids. The amount and types of the gases used in the experiment were ideally determined to allow amino acids to originate. The amount of energy supplied to the system was neither too much nor too little, but arranged precisely to enable the necessary reactions to occur. The experimental apparatus was isolated, so that it would not allow the leaking of any harmful, destructive, or any other kind of elements to hinder the formation of amino acids. No elements, minerals or compounds that were likely to have been present on the primordial earth, but which would have changed the course of the reactions, were included in the experiment. Oxygen, which would have prevented the formation of amino acids because of oxidation, is only one of these destructive elements. Even under such ideal laboratory conditions, it was impossible for the amino acids produced to survive and avoid destruction without the "cold trap" mechanism.

In fact, by his experiment, Miller destroyed evolution's claim that "life emerged as the result of unconscious coincidences". That is because, if the experiment proves anything, it is that amino acids can only be produced in a controlled laboratory environment where all the conditions are specifically designed by conscious intervention. That is, the power that brings about life cannot be by unconscious chance but rather by conscious creation.

The reason evolutionists do not accept this evident reality is their blind adherence to prejudices that are totally unscientific. Interestingly enough, **Harold Urey**, who organised the Miller experiment with his student Stanley Miller, made the following confession on the subject:

> All of us who study the origin of life find that the more we look into it, **the more we feel it is too complex to have evolved anywhere**. We all believe as an article of faith that life evolved from dead matter on this planet. It is just that its complexity is so great, it is hard for us to imagine that it did.[120]

Primordial World Atmosphere and Proteins

Evolutionist sources use the Miller experiment, despite all of its inconsistencies, to try to gloss over the question of the origin of amino acids. By giving the impression that the issue has long since been resolved by that invalid experiment, they try to paper over the cracks in the theory of evolution.

However, to explain the second stage of the origin of life, evolutionists faced an even greater problem than that of the formation of amino acids-namel, the origin of **proteins**, the building blocks of life, which are composed of hundreds of different amino acids bonding with each other in a particular order.

Claiming that proteins were formed by chance under natural conditions is even more unrealistic and unreasonable than claiming that amino acids were formed by chance. In the preceding pages we have seen the mathematical impossibility of the haphazard uniting of amino acids in proper sequences to form proteins with probability calculations. Now, we will examine the impossibility of proteins being produced chemically under primordial earth conditions.

Protein Synthesis is not Possible in Water

As we saw before, when combining to form proteins, amino acids form a special bond with one another called the "peptide bond". A water molecule is released during the formation of this peptide bond.

This fact definitely refutes the evolutionist explanation that primordial life originated in water, because according to the "**Le Châtelier principle**" in chemistry, it is not possible for a reaction that releases water (a condensation reaction) to take place in a hydrous environment. The chances of this kind of a reaction happening in a hydrate environment is said to "have the least probability of occurring" of all chemical reactions.

Hence the ocean, which is claimed to be where life began and amino acids originated, is definitely not an appropriate setting for amino acids to form proteins. On the other hand, it would be irrational for evolutionists to change their minds and claim that life originated on land, because the only environment where amino acids could have been protected from ultraviolet radiation is in the oceans and seas. On land, they would be destroyed by ultraviolet rays. The Le Châtelier Principle disproves the claim of the formation of life in the sea. This is another dilemma confronting evolution.

Another Desperate Effort: Fox's Experiment

Challenged by the above dilemma, evolutionists began to invent unrealistic scenarios based on this "water problem" that so definitively refuted their theories. Sydney Fox was one of the best known of these researchers. Fox advanced the following theory to solve this problem. According to him, the first amino acids must have been transported to some cliffs near a volcano right after their formation in the primordial ocean. The water contained in this mixture that included the amino acids present on the cliffs, must have evaporated when the temperature increased above boiling point. The amino acids which were "dried out" in this way, could

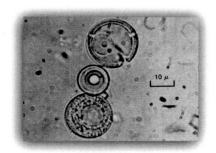

In his experiment, Fox produced a substance called "proteinoid". Proteinoids were randomly assembled combinations of amino acids. Unlike proteins of living things, these were useless and non-functional chemicals. Here is an electron microscope vision of proteinoid particles.

then have combined to form proteins.

However this "complicated" way out was not accepted by many people in the field, because the amino acids could not have endured such high temperatures. Research confirmed that amino acids are immediately destroyed at very high temperatures.

But Fox did not give up. He combined purified amino acids in the laboratory, "under very special conditions" by heating them in a dry environment. The amino acids combined, but still no proteins were obtained. What he actually ended up with was simple and disordered loops of amino acids, arbitrarily combined with each other, and these loops were far from resembling any living protein. Furthermore, if Fox had kept the amino acids at a steady temperature, then these useless loops would also have disintegrated.[121]

Another point that nullified the experiment was that Fox did not usethe useless end products obtained in Miller's experiment; rather, he used pure amino acids from living organisms. This experiment, however, which was intended to be a continuation of Miller's experiment, should have started out from the results obtained by Miller. Yet neither Fox, nor any other researcher, used the useless amino acids Miller produced.[122]

Fox's experiment was not even welcomed in evolutionist circles, because it was clear that the meaningless amino acid chains that he obtained (which he termed "proteinoids") could not have formed under natural conditions. Moreover, proteins, the basic units of life, still could not be produced. The problem of the origin of proteins remained unsolved. In an article in the popular science magazine, *Chemical Engineering News*, which appeared in the 1970s, Fox's experiment was mentioned as follows:

> Sydney Fox and the other researchers managed to unite the amino acids in the shape of "proteinoids" by using very special heating techniques under

INANIMATE MATTER CANNOT GENERATE LIFE

A number of evolutionist experiments such as the Miller Experiment and the Fox Experiment have been devised to prove the claim that inanimate matter can organise itself and generate a complex living being. This is an utterly unscientific conviction: every observation and experiment has incontrovertibly proven that matter has no such ability. The famous English astronomer and mathematician Sir Fred Hoyle notes that matter cannot generate life by itself, without deliberate interference:

> If there were a basic principle of matter which somehow drove organic systems toward life, its existence should easily be demonstrable in the laboratory. One could, for instance, take a swimming bath to represent the primordial soup. Fill it with any chemicals of a non-biological nature you please. Pump any gases over it, or through it, you please, and shine any kind of radiation on it that takes your fancy. Let the experiment proceed for a year and see how many of those 2,000 enzymes (proteins produced by living cells) have appeared in the bath. I will give the answer, and so save the time and trouble and expense of actually doing the experiment. You will find nothing at all, except possibly for a tarry sludge composed of amino acids and other simple organic chemicals.[1]

Evolutionist biologist Andrew Scott admits the same fact:

> Take some matter, heat while stirring and wait. That is the modern version of Genesis. The 'fundamental' forces of gravity, electromagnetism and the strong and weak nuclear forces are presumed to have done the rest... But how much of this neat tale is firmly established, and how much remains hopeful speculation? In truth, the mechanism of almost every major step, from chemical precursors up to the first recognizable cells, is the subject of either controversy or complete bewilderment.[2]

1- Fred Hoyle, The Intelligent Universe, New York, Holt, Rinehard & Winston, 1983, p. 256
2- Andrew Scott, "Update on Genesis", New Scientist, vol. 106, May 2nd, 1985, p. 30

conditions which in fact did not exist at all in the primordial stages of Earth. Also, they are not at all similar to the very regular proteins present in living things. They are nothing but useless, irregular chemical stains. It was explained that even if such molecules had formed in the early ages, they would definitely be destroyed.[123]

Indeed, the proteinoids Fox obtained were totally different from real proteins both in structure and function. The difference between proteins and these proteinoids was as huge as the difference between a piece of high-tech equipment and a heap of unprocessed iron.

Furthermore, there was no chance that even these irregular amino acid chains could have survived in the primordial atmosphere. Harmful and destructive physical and chemical effects caused by heavy exposure to ultraviolet light and other unstable natural conditions would have caused these proteinoids to disintegrate. Because of the Le Châtelier principle, it was also impossible for the amino acids to combine underwater, where ultraviolet rays would not reach them. In view of this, the idea that the proteinoids were the basis of life eventually lost support among scientists.

The Miraculous Molecule: DNA

Our examinations so far have shown that the theory of evolution is in a serious quandary at the molecular level. Evolutionists have shed no light on the formation of amino acids at all. The formation of proteins, on the other hand, is another mystery all its own.

Yet the problems are not even limited just to amino acids and proteins: These are only the beginning. Beyond them, the extremely complex structure of the cell leads evolutionists to yet another impasse. The reason for this is that the cell is not just a heap of amino-acid-structured proteins, but rather the most complex system man has ever encountered.

While the theory of evolution was having such trouble providing a coherent explanation for the existence of the molecules that are the basis of the cell structure, developments in the science of genetics and the discovery of nucleic acids (DNA and RNA) produced brand-new problems for the theory. In 1953, James Watson and Francis Crick launched a new age in biology with their work on the structure of DNA. Soon, many scientists were directing their attention to the science of genetics. Today, after many years of research, the structure of DNA has been to a great extent unravelled.

The molecule known as DNA, which is found in the nucleus of each of the 100 trillion cells in our bodies, contains the complete blueprint for the construction of the human body. The information regarding all the characteristics of a person, from physical appearance to the structure of the inner organs, is recorded in DNA within the sequence of four special bases that make up the giant molecule. These bases are known as A, T, G, and C, according to the initial letters of their names. All the structural differences among people depend on variations in the sequences of these letters. This is a sort of a data-bank composed of four letters.

The sequential order of the letters in DNA determines the structure of a human being down to its slightest details. In addition to features such as

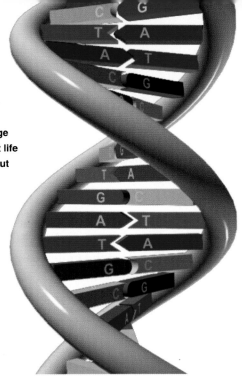

All information about living beings is stored in the DNA molecule. This incredibly efficient information storage method alone is a clear evidence that life did not come into being by chance, but has been purposefully designed, or, better to say, marvellously created.

height, and eye, hair and skin colours, the DNA in a single cell also contains the design of the 206 bones, the 600 muscles, the 100 billion nerve cells (neurons), 1.000 trillion connections between the neurons of the brain, 97,000 kilometres of veins, and the 100 trillion cells of the human body. **If we were to write down the information coded in DNA, then we would have to compile a giant library consisting of 900 volumes of 500 pages each**. But the information this enormous library would hold is encoded inside the DNA molecules in the cell nucleus, which is far smaller than the 1/100th-of-a-millimetre-long cell itself.

Can DNA Come into Being by Chance?

At this point, there is an important detail that deserves attention. An error in the sequence of the nucleotides making up a gene would render that gene completely useless. When it is considered that there are 200,000 genes in the human body, it becomes clearer how impossible it is for the millions of nucleotides making up these genes to have been formed, in the right sequence, by chance. The evolutionist biologist Frank Salisbury has comments on this impossibility:

> A medium protein might include about 300 amino acids. The DNA gene controlling this would have about 1,000 nucleotides in its chain. Since there are four kinds of nucleotides in a DNA chain, one consisting of 1,000 links could exist in $4^{1,000}$ forms. Using a little algebra (logarithms) we can see that $4^{1000}=10^{600}$. Ten multiplied by itself 600 times gives the figure 1 followed by 600 zeros! This number is completely beyond our comprehension.[124]

The number 4^{1000} is the equivalent of 10^{600}. This means 1 followed by

The Molecular Impasse of Evolution

600 zeros. As 1 with 12 zeros after it indicates a trillion, 600 zeros represents an inconceivable number. The impossibility of the formation of RNA and DNA by a coincidental accumulation of nucleotides is expressed by the French scientist Paul Auger in this way:

> We have to sharply distinguish the two stages in the chance formation of complex molecules such as nucleotides by chemical events. The production of nucleotides one by one - which is possible- and the combination of these with in very special sequences. The second is absolutely impossible.[125]

For many years, Francis Crick believed in the theory of molecular evolution, but eventually even he had to admit to himself that such a complex molecule could not have emerged spontaneously by coincidence, as the result of an evolutionary process:

> An honest man, armed with all the knowledge available to us now, could only state that, in some sense, the origin of life appears at the moment to be almost a miracle.[126]

The Turkish evolutionist Professor Ali Demirsoy was forced to make the following confession on the issue:

> In fact, the probability of the formation of a protein and a nucleic acid (DNA-RNA) is a probability way beyond estimating. Furthermore, the chance of the emergence of a certain protein chain is so slight as to be called astronomic.[127]

A very interesting paradox emerges at this point: While DNA can only replicate with the help of special proteins (enzymes), the synthesis of these proteins can only be realised by the information encoded in DNA. As

Watson and Crick with a stick model of the DNA molecule.

they both depend on each other, either they have to exist at the same time for replication, or one of them has to be "created" before the other. The American microbiologist Homer Jacobson comments:

> Directions for the reproduction of plans, for energy and the extraction of parts from the current environment, for the growth sequence, and for the effector mechanism translating instructions into growth-*all* had to be simultaneously present at that moment [when life began]. This combination of events has seemed an incredibly unlikely happenstance, and has often been ascribed to divine intervention.[128]

The quotation above was written two years after the discovery of the structure of DNA by Watson and Crick. But despite all the developments in science, this problem for evolutionists remains unsolved. Two German scientists Junker and Scherer explained that the synthesis of each of the molecules required for chemical evolution, necessitates distinct conditions, and that the probability of the compounding of these materials having theoretically very different acquirement methods is zero:

> Until now, no experiment is known in which we can obtain all the molecules necessary for chemical evolution. Therefore, it is essential to produce various molecules in different places under very suitable conditions and then to carry them to another place for reaction by protecting them from harmful elements like hydrolysis and photolysis.[129]

In short, the theory of evolution is unable to prove any of the evolutionary stages that allegedly occur at the molecular level. Rather than providing answers to such questions, the progress of science renders them even more complex and inextricable.

Interestingly enough, most evolutionists believe in this and similar totally unscientific fairy tales as if they were true, because accepting intelligent design means accepting creation-and they have conditioned themselves not to accept this truth. One famous biologist from Australia, Michael Denton, discusses the subject in his book *Evolution: A Theory in Crisis*:

Prof. Francis Crick: "The origin of life appears to be almost a miracle."

> To the skeptic, the proposition that the genetic programmes of higher organisms, consisting of something close to a thousand million bits of information, equivalent to the sequence of letters in a small library of 1,000 volumes, containing in encoded form countless thousands of intricate algorithms control-

ling, specifying, and ordering the growth and development of billions and billions of cells into the form of a complex organism, were composed by **a purely random process is simply an affront to reason. But to the Darwinist, the idea is accepted without a ripple of doubt-the paradigm takes precedence!**[130]

Another Evolutionist Vain Attempt: "The RNA World"

The discovery in the 1970s that the gasses originally existing in the primitive atmosphere of the earth would have rendered amino acid synthesis impossible was a serious blow to the theory of molecular evolution. Evolutionists then had to face the fact that the "primitive atmosphere experiments" by Stanley Miller, Sydney Fox, Cyril Ponnamperuma and others were invalid. For this reason, in the 1980s the evolutionists tried again. As a result, the "RNA World" hypothesis was advanced. This scenario proposed that, not proteins, but rather the RNA molecules that contained the information for proteins, were formed first.

According to this scenario, advanced by Harvard chemist Walter Gilbert in 1986, based on a discovery about "ribozymes" by Thomas Cech , billions of years ago an RNA molecule capable of replicating itself formed somehow by accident. Then this RNA molecule started to produce proteins, having been activated by external influences. Thereafter, it became necessary to store this information in a second molecule, and somehow the DNA molecule emerged to do that.

Made up as it is of a chain of impossibilities in each and every stage, this scarcely credible scenario, far from providing any explanation of the origin of life, only magnified the problem, and raised many unanswerable questions:

1. Since it is impossible to accept the coincidental formation of even one of the nucleotides making up RNA, how can it be possible for these imaginary nucleotides to form RNA by coming together in a particular sequence? Evolutionist John Horgan admits the impossibility of the chance formation of RNA;

> As researchers continue to examine the RNA-world concept closely, more problems emerge. How did RNA initially arise? RNA and its components are difficult to synthesize in a laboratory under the best of conditions, much less under really plausible ones.[131]

2. Even if we suppose that it formed by chance, how could this RNA,

CONFESSIONS FROM EVOLUTIONISTS

Probabilistic calculations make it clear that complex molecules such as proteins and nucleic acids (RNA and DNA) could not ever have been formed by chance independently of each other. Yet evolutionists have to face the even greater problem that all these complex molecules have to coexist simultaneously in order for life to exist at all. Evolutionary theory is utterly confounded by this requirement. This is a point on which some leading evolutionists have been forced to confession. For instance, Stanley Miller's and Francis Crick's close associate from the University of San Diego California, reputable evolutionist Dr. Leslie Orgel says:

> It is extremely improbable that proteins and nucleic acids, both of which are structurally complex, arose spontaneously in the same place at the same time. Yet it also seems impossible to have one without the other. And so, at first glance, one might have to conclude that life could never, in fact, have originated by chemical means.[1]

The same fact is also admitted by other scientists:

> DNA cannot do its work, including forming more DNA, without the help of catalytic proteins, or enzymes. In short, proteins cannot form without DNA, but neither can DNA form without proteins.[2]

> How did the Genetic Code, along with the mechanisms for its translation (ribosomes and RNA molecules), originate? For the moment, we will have to content ourselves with a sense of wonder and awe, rather than with an answer.[3]

The *New York Times* science correspondent, Nicholas Wade made this comment in an article dated 2000:

> Everything about the origin of life on Earth is a mystery, and it seems the more that is known, the more acute the puzzle get.[4]

1- Leslie E. Orgel, "The Origin of Life on Earth", *Scientific American*, vol. 271, October 1994, p. 78
2- John Horgan, "In the Beginning", Scientific American, vol. 264, February 1991, p. 119
3- Douglas R. Hofstadter, Gödel, Escher, Bach: An Eternal Golden Braid, New York, Vintage Books, 1980, p. 548
4- Nicholas Wade, "Life's Origins Get Murkier and Messier", The New York Times, June 13, 2000, pp. D1-D2

consisting of just a nucleotide chain, have "decided" to self-replicate, and with what kind of mechanism could it have carried out this self-replicating process? Where did it find the nucleotides it used while self-replicating? Even evolutionist microbiologists Gerald Joyce and Leslie Orgel express the desperate nature of the situation in their book *In the RNA World*:

> This discussion... has, in a sense, focused on a straw man: the myth of a self-replicating RNA molecule that arose de novo from a soup of random polynucleotides. Not only is such a notion unrealistic in light of our current understanding of prebiotic chemistry, but it would strain the credulity of even an optimist's view of RNA's catalytic potential.[132]

3. Even if we suppose that there was self-replicating RNA in the primordial world, that numerous amino acids of every type ready to be used by RNA were available, and that all of these impossibilities somehow took place the situation still does not lead to the formation of even one single protein. For RNA only includes information concerning the structure of proteins. Amino acids, on the other hand, are raw materials. Nevertheless, there is no mechanism for the production of proteins. To consider the existence of RNA sufficient for protein production is as nonsensical as expecting a car to assemble itself simplyh throwing the blueprint onto a heap of parts piled up on top of each other. A blueprint cannot produce a car all by itself without a factory and workers to assemble the parts according to the instructions contained in the blueprint; in the same way, the blueprint contained in RNA cannot produce proteins by itself without the cooperation of other cellular components which follow the instructions contained in the RNA.

Proteins are produced in the ribosome factory with the help of many enzymes and as a result of extremely complex processes within the cell. The ribosome is a complex cell organelle made up of proteins. This leads, therefore, to another unreasonable supposition-that ribosomes, too, should have come into existence by chance at the same time. Even Nobel Prize winner Jacques Monod, who was one of the most fanatical defenders of evolution- and atheism-explained that protein synthesis can by no means be considered to depend merely on the information in the nucleic acids:

> The code is meaningless unless translated. The modern cell's translating machinery consists of at least 50 macromolecular components, *which are themselves coded in DNA: the code cannot be translated otherwise than by products of translation themselves. It is the modern expression of omne vivum ex ovo.* When and how did this circle become closed? It is exceedingly difficult to imagine.[133]

How could an RNA chain in the primordial world have taken such a decision, and what methods could it have employed to make protein production happen by doing the work of 50 specialized particles on its own? Evolutionists have no answer to these questions.

Dr. Leslie Orgel, one of the associates of Stanley Miller and Francis Crick from the University of California at San Diego, uses the term "scenario" for the possibility of "the origination of life through the RNA World". Orgel described what kind of features this RNA have had to have and how impossible this would have been in his article "The Origin of Life" published in *American Scientist* in October 1994:

This scenario could have occured, we noted, if prebiotic RNA had two properties not evident today: A capacity to replicate without the help of proteins and an ability to catalyze every step of protein synthesis.[134]

As should by now be clear, to expect these two complex and extremely essential processes from a molecule such as RNA is only possible from the evolutionist's viewpoint and with the help of his power of imagination. Concrete scientific facts, on the other hand, makes it explicit that the RNA World hypothesis, which is a new model proposed for the chance formation of life, is an equally implausible fable.

Biochemist Gordon C. Mills from the University of Texas and Molecular biologist Dean Kenyon from San Francisco State University assess the flaws of the RNA World scenario, and reach to a brief conclusion in their article titled " The RNA World: A Critique": *"RNA is a remarkable molecule. The RNA World hypothesis is another matter. We see no grounds for considering it established, or even promising."* [135]

Science writer Brig Klyce's 2001 article explains that evolutionist scientists are very persistent on this issue, but the results obtained so far have already shown that these efforts are all in vain:

Research in the RNA world is a medium-sized industry. This research has demonstrated how exceedingly difficult it would be for living cells to originate by chance from nonliving matter in the time available on Earth. That demonstration is a valuable contribution to science. Additional research will be valuable as well. But to keep insisting that life can spontaneously emerge from nonliving chemicals in the face of the newly comprehended difficulties is puzzling. It is reminiscent of the work of medieval alchemists who persistently tried to turn lead into gold.[136]

Life is a Concept Beyond Mere Heaps of Molecules

So far, we have examined how impossible the accidental formation of life is. Let us again ignore these impossibilities for just a moment. Let us suppose that a protein molecule was formed in the most inappropriate, most uncontrolled environment such as the primordial earth conditions. The formation of only one protein would not be sufficient; this protein would have to wait patiently for thousands, maybe millions of years in this uncontrolled environment without sustaining any damage, until another molecule was formed beside it by chance under the same conditions. It would have to wait until millions of correct and essential proteins were formed side by side in the same setting all "by chance". Those that formed earlier had to be patient enough to wait, without being destroyed despite ultraviolet rays and harsh mechanical effects, for the others to be formed right next to them. Then these proteins in adequate number, which all originated at the very same spot, would have to come together by making meaningful combinations and form the organelles of the cell. No extraneous material, harmful molecule, or useless protein chain may interfere with them. Then, even if these organelles were to come together in an extremely harmonious and co-operative way within a plan and order, they must take all the necessary enzymes beside themselves and become covered with a membrane, the inside of which must be filled with a special liquid to prepare the ideal environment for them. Now even if all these "highly unlikely" events actually occurred by chance, would this molecular heap come to life?

The answer is No, because research has revealed that **the mere combination of all the materials essential for life is not enough for life to get started**. Even if all the essential proteins for life were collected and put in a test tube, these efforts would not result with producing a living cell. All the experiments conducted on this subject have proved to be unsuccessful. All observations and experiments indicate that life can only originate from life. The assertion that life evolved from non-living things, in other words, "abiogenesis", is a tale only existing in the dreams of the evolutionists and completely at variance with the results of every experiment and observation.

In this respect, the first life on earth must also have originated from other life. This is a reflection of God's epithet of "Hayy" (The Owner of Life). Life can only start, continue, and end by His will. As for evolution, not only is it unable to explain how life began, it is also unable to explain how the materials essential for life have formed and come together.

Chandra Wickramasinghe describes the reality he faced as a scientist who had been told throughout his life that life had emerged as a result of chance coincidences:

> From my earliest training as a scientist, I was very strongly brainwashed to believe that science cannot be consistent with any kind of deliberate creation. That notion has had to be painfully shed. At the moment, I can't find any rational argument to knock down the view which argues for conversion to God. We used to have an open mind; now we realize that the only logical answer to life is creation-and not accidental random shuffling.[137]

CHAPTER 11

Thermodynamics Falsifies Evolution

The second law of thermodynamics, which is accepted as one of the basic laws of physics, holds that under normal conditions all systems left on their own tend to become disordered, dispersed, and corrupted in direct relation to the amount of time that passes. Everything, whether living or not wears out, deteriorates, decays, disintegrates, and is destroyed. This is the absolute end that all beings will face one way or another, and according to the law, the process cannot be avoided.

This is something that all of us have observed. For example if you take a car to a desert and leave it there, you would hardly expect to find it in a better condition when you came back years later. On the contrary, you would see that its tires had gone flat, its windows had been broken, its chassis had rusted, and its engine had stopped working. The same inevitable process holds true for living things.

The second law of thermodynamics is the means by which this natural process is defined with physical equations and calculations.

This famous law of physics is also known as "the law of entropy". In physics, entropy is the measure of the disorder of a system. A system's entropy increases as it moves from an ordered, organised, and planned state towards a more disordered, dispersed, and unplanned one. The more disorder there is in a system, the higher its entropy is. The law of entropy holds that the entire universe is unavoidably proceeding towards a more disordered, unplanned, and disorganised state.

The truth of the second law of thermodynamics, or the law of entropy, has been experimentally and theoretically established. All foremost scientists agree that the law of entropy will remain the principle paradigm for the foreseeable future. Albert Einstein, the greatest scientist of our age, described it as the "premier law of all of science". Sir Arthur Eddington also referred to it as the "supreme metaphysical law of the entire universe".[138]

Evolutionary theory ignores this fundamental law of physics. The mechanism offered by evolution totally contradicts the second law. The theory of evolution says that disordered, dispersed, and lifeless atoms and molecules spontaneously came together over time, in a particular order, to form extremely complex molecules such as proteins, DNA, and RNA, whereupon millions of different living species with even more complex structures gradually emerged. According to the theory of evolution, this supposed process-which yields a more planned, more ordered, more complex and more organised structure at each stage-was formed all by itself under natural conditions. The law of entropy makes it clear that this so-called natural process utterly contradicts the laws of physics.

Evolutionist scientists are also aware of this fact. J.H. Rush states:

> In the complex course of its evolution, life exhibits a remarkable contrast to the tendency expressed in the Second Law of Thermodynamics. Where the Second Law expresses an irreversible progression toward increased entropy and disorder, life evolves continually higher levels of order.[139]

> The law of thermodynamics holds that natural conditions always lead to disorder and loss of information. Evolutionary theory, on the other hand, is an unscientific belief that utterly contradicts with this law.

The evolutionist author Roger Lewin expresses the thermodynamic impasse of evolution in an article in Science:

> One problem biologists have faced is the apparent contradiction by evolution of the second law of thermodynamics. Systems should decay through time, giving less, not more, order.[140]

Another defender of the theory of evolution, George Stravropoulos states the thermodynamic impossibility of the spontaneous formation of life and the impossibility of explaining the existence of complex living mechanisms by natural laws in the well-known evolutionist journal *American Scientist*:

> Yet, under ordinary conditions, no complex organic molecule can ever form spontaneously but will rather disintegrate, in agreement with the second law. Indeed, the more complex it is, the more unstable it will be, and the more assured, sooner or later, its disintegration. Photosynthesis and all life processes, and even life itself, *cannot* yet be understood in terms of thermodynamics or any other exact science, despite the use of confused or deliberately confusing language.[141]

As we have seen, the second law of thermodynamics constitutes an insurmountable obstacle for the scenario of evolution, in terms of both science and logic. Unable to offer any scientific and consistent explanation to overcome this obstacle, evolutionists can only do so in their imagination. For instance, the well-known evolutionist Jeremy Rifkin notes his belief that evolution overwhelms this law of physics with a "magical power":

> The Entropy Law says that evolution dissipates the overall available energy for life on this planet. Our concept of evolution is the exact opposite. We believe that evolution somehow magically creates greater overall value and order on earth.[142]

These words well indicate that evolution is a dogmatic belief rather than a scientific thesis.

The Myth of the "Open System"

Some proponents of evolution have recourse to an argument that the second law of thermodynamics holds true only for "closed systems", and that "open systems" are beyond the scope of this law.

An "open system" is a thermodynamic system in which energy and matter flow in and out. Evolutionists hold that the world is an open system: that it is constantly exposed to an energy flow from the sun, that the law of entropy does not apply to the world as a whole, and that ordered, complex living beings can be generated from disordered, simple, and inanimate structures.

However, there is an obvious distortion here. **The fact that a system has an energy inflow is not enough to make that system ordered. Specific mechanisms are needed to make the energy functional.** For instance, a car needs an engine, a transmission system, and related control mechanisms to convert the energy in petrol to work. Without such an energy conversion system, the car will not be able to use the energy stored in petrol.

The same thing applies in the case of life as well. It is true that life derives its energy from the sun. However, solar energy can only be converted into chemical energy by the incredibly complex energy conversion systems in living things (such as photosynthesis in plants and the digestive systems of humans and animals). No living thing can live without such energy conversion systems. Without an energy conversion system, the sun is nothing but a source of destructive energy that burns, parches, or melts.

As may be seen, a thermodynamic system without an energy conversion mechanism of some sort is not advantageous for evolution, be it open or closed. No one asserts that such complex and conscious mechanisms could have existed in nature under the conditions of the primeval earth. Indeed, the real problem confronting evolutionists is the question of how complex energy-converting mechanisms such as photosynthesis in plants, which cannot be duplicated even with modern technology, could have come into being on their own.

The influx of solar energy into the world would be unable to bring about order on its own. Moreover, no matter how high the temperature may become, amino acids resist forming bonds in ordered sequences. Energy by itself is incapable of making amino acids form the much more complex molecules of proteins, or of making proteins from the much complex and organised structures of cell organelles. The real and essential source of this organisation at all levels is intelligent design: in a word, creation.

The Myth of the "Self Organization of Matter"

Quite aware that the second law of thermodynamics renders evolution impossible, some evolutionist scientists have made speculative attempts to square the circle between the two, in order to be able to claim that evolution is possible. As usual, even those endeavours show that the theory of evolution faces an inescapable impasse.

One person distinguished by his efforts to marry thermodynamics and evolution is the Belgian scientist Ilya Prigogine. Starting out from chaos theory, Prigogine proposed a number of hypotheses in which order develops from chaos (disorder). He argued that some open systems can portray a decrease in entropy due to an influx of outer energy and the outcoming "ordering" is a proof that "matter can organise itself." Since then, the concept of the "self-organization of matter" has been quite popular among evolutionists and materialists. They act like they have found a materialistic origin for the complexity of life and a materialistic solution for the problem of life's origin.

But a closer look reveals that this argument is totally abstract and in fact just wishful thinking. Moreover, it includes a very naive deception. The deception lies in the deliberate confusing of two distinct concepts, **"ordered"** and **"organised."**[143]

We can make this clear with an example. Imagine a completely flat beach on the seashore. When a strong wave hits the beach, mounds of sand, large and small, form bumps on the surface of the sand.

This is a process of "ordering": The seashore is an open system and the energy flow (the wave) that enters it can form simple patterns in the sand, which look completely regular. From the thermodynamic point of view, it can set up order here where before there was none. But we must make it clear that those same waves cannot build a castle on the beach. If we see a castle there, we are in no doubt that someone has constructed it, because the castle is an "organised" system. In other words, it possesses a clear design and information. Every part of it has been made by a conscious entity in a planned manner.

The difference between the sand and the castle is that the former is an organised complexity, whereas the latter possesses only order, brought about by simple repetitions. The order formed from repetitions is as if an object (in other words the flow of energy entering the system) had fallen on the letter "a" on a typewriter keyboard, writing "aaaaaaaaaaaaaaaa" hundreds of times. But the string of "a"s in an order repeated in this manner contains no information, and no complexity. In order to write a complex chain of letters actually containing information (in other words a meaningful sequence, paragraph or book), the presence of intelligence is essential.

The same thing applies when wind blows into a dusty room. When the wind blows in, the dust which had been lying in an even layer may gather in one corner of the room. This is also a more ordered situation than that which existed before, in the thermodynamic sense, but the individual specks of dust cannot form a portrait of someone on the floor in an organised manner.

This means that complex, organised systems can never come about as the result of natural processes. Although simple examples of order can happen from time to time, these cannot go beyond limits.

But evolutionists point to this self-ordering which emerges through natural processes as a most important proof of evolution, portray such cases as examples of "self-organization". As a result of this confusion of concepts, they propose that living systems could develop their own accord from occurrences in nature and chemical reactions. The methods and studies employed by Prigogine and his followers, which we considered above, are based on this deceptive logic.

The American scientists Charles B. Thaxton, Walter L. Bradley and Roger L. Olsen, in their book titled *The Mystery of Life's Origin*, explain this fact as follows:

> ...In each case random movements of molecules in a fluid are spontaneously replaced by a highly ordered behavior. Prigogine, Eigen, and others have

suggested that a similar sort of self-organization may be intrinsic in organic chemistry and can potentially account for the highly complex macromolecules essential for living systems. But such analogies have scant relevance to the origin-of-life question. A major reason is that they fail to distinguish between order and complexity... Regularity or order cannot serve to store the large amount of information required by living systems. A highly irregular, but specified, structure is required rather than an ordered structure. This is a serious flaw in the analogy offered. There is no apparent connection between the kind of spontaneous ordering that occurs from energy flow through such systems and the work required to build aperiodic information-intensive macromolecules like DNA and protein.[144]

In fact even Prigogine himself has accepted that the theories he has produced for the molecular level do not apply to living systems-for instance, a living cell:

The problem of biological order involves the transition from the molecular activity to the supermolecular order of the cell. This problem is far from being solved.[145]

So why do evolutionists continue to believe in scenarios such as the "self organization of matter", which have no scientific foundation? Why are they so determined to reject the intelligence and planning that so clearly can be seen in living systems? The answer is that they have a dogmatic faith in materialism and they believe that matter has some mysterious power to create life. A professor of chemistry from New York University and DNA expert, Robert Shapiro, explains this belief of evolutionists about the "self-organization of matter" and the materialist dogma lying at its heart as follows:

Another evolutionary principle is therefore needed to take us across the gap from mixtures of simple natural chemicals to the first effective replicator. This principle has not yet been described in detail or demonstrated, but it is anticipated, and given names such as chemical evolution and self-organization of matter. The existence of the principle is taken for granted in the philosophy of dialectical materialism, as applied to the origin of life by Alexander Oparin.[146]

All this situation clearly demonstrates that evolution is a dogma that is against emprical science and the origin of living beings can only be explained by the intervention of a supernatural power. That supernatural power is the creation of God, who created the entire universe from nothing. Science has proven that evolution is still impossible as far as thermodynamics is concerned and the existence of life has no explanation but Creation.

CHAPTER 12

Design and Coincidence

In the previous chapter, we have examined how impossible the accidental formation of life is. Let us again ignore these impossibilities for just a moment. Let us suppose that millions of years ago a cell was formed which had acquired everything necessary for life, and that it duly "came to life". Evolution again collapses at this point. For even if this cell had existed for a while, it would eventually have died and after its death, nothing would have remained, and everything would have reverted to where it had started. This is because this first living cell, lacking any genetic information, would not have been able to reproduce and start a new generation. Life would have ended with its death.

The genetic system does not only consist of DNA. The following things must also exist in the same environment: enzymes to read the code on the DNA, messenger RNA to be produced after reading these codes, a ribosome to which messenger RNA will attach according to this code, transfer RNA to transfer the amino acids to the ribosome for use in production, and extremely complex enzymes to carry out numerous intermediary processes. Such an environment cannot exist anywhere apart from aa totally isolated and completely controlled environment such as the cell, where all the essential raw materials and energy resources exist.

As a result, organic matter can self-reproduce only if it exists as a fully developed cell with all its organelles and in an appropriate environment where it can survive, exchange materials, and get energy from its surroundings. This means that the first cell on earth was formed "all of a sudden" together with its incredibly complex structure.

So, **if a complex structure came into existence all of a sudden, what does this mean?**

Let us ask this question with an example. Let us liken the cell to a high-tech car in terms of its complexity. (In fact, the cell is a much more complex and developed system than a car with its engine and all its technical equipment.) Now let us ask the following question: What would you think if you went out hiking in the depths of a thick forest and ran across a brand-new car among the trees? Would you imagine that various elements in the forest had come together by chance over millions of years and produced such a vehicle? All the parts in the car are made of products such as iron, copper, and rubber-the raw ingredients for which are all found on the earth-but would this fact lead you to think that these materials had synthesised "by chance" and then come together and manufactured such a car?

There is no doubt that anyone with a sound mind would realise that the car was the product of an intelligent design-in other words, a factory-and wonder what it was doing there in the middle of the forest. The sudden emergence of a complex structure in a complete form, quite out of the blue, shows that this is the work of an intelligent agent. A complex system like the cell is no doubt created by a superior will and wisdom. In other words, it came into existence as a creation of God.

Believing that pure chance can produce perfect designs goes well beyond the bounds of reason. Yet, every "explanation put forward by the theory of evolution regarding the origin of life is like that. One outspoken authority on this issue is the famous French zoologist Pierre-Paul Grassé, the former president of the French Academy of Sciences. Grassé is a materialist, yet he acknowledges that Darwinist theory is unable to explain life and makes a point about the logic of "coincidence", which is the backbone of Darwinism:

> The opportune appearance of mutations permitting animals and plants to meet their needs seems hard to believe. Yet the Darwinian theory is even more demanding: A single plant, a single animal would require thousands and thousands of lucky, appropriate events. Thus, miracles would become the rule: events with an infinitesimal probability could not fail to occur... **There is no law against daydreaming, but science must not indulge in it.**[147]

Grasse summarises what the concept of "coincidence" means for evolutionists: "...**Chance becomes a sort of providence, which, under the cover of atheism, is not named but which is secretly worshipped.**"[148]

The logical failure of evolutionists is an outcome of their enshrining the concept of coincidence. In the Qur'an, it is written that those who worship beings other than God are devoid of understanding;

> **They have hearts wherewith they understand not, eyes wherewith they see not, and ears wherewith they hear not. They are like cattle - nay more misguided: for they are heedless (of warning).** (Surat al-Araf : 179)

Darwinian Formula!

Besides all the technical evidence we have dealt with so far, let us now for once, examine what kind of a superstition the evolutionists have with an example so simple as to be understood even by children:

Evolutionary theory asserts that life is formed by chance. According to this claim, lifeless and unconscious atoms came together to form the cell and then they somehow formed other living things, including man. Let us think about that. When we bring together the elements that are the building-blocks of life such as carbon, phosphorus, nitrogen and potassium, only a heap is formed. No matter what treatments it undergoes, this atomic heap cannot form even a single living being. If you like, let us formulate an "experiment" on this subject and let us examine on the behalf of evolutionists what they really claim without pronouncing loudly under the name "Darwinian formula":

Let evolutionists put plenty of materials present in the composition of living beings such as phosphorus, nitrogen, carbon, oxygen, iron, and magnesium into big barrels. Moreover, let them add in these barrels any material that does not exist under normal conditions, but they think as necessary. Let them add in this mixture as many amino acids-which have no possibility of forming under natural conditions-and as many proteins-a single one of which has a formation probability of 10^{-950}-as they like. Let them expose these mixtures to as much heat and moisture as they like. Let them stir these with whatever technologically developed device they like. Let them put the foremost scientists beside these barrels. Let these experts wait in turn beside these barrels for billions, and even trillions of years. Let them be free to use all kinds of conditions they believe to be necessary for a human's formation. No matter what they do, they cannot produce from these barrels a human, say a professor that examines his cell structure under the electron microscope. They cannot produce giraffes, lions, bees, canaries, horses, dolphins, roses, orchids, lilies, carnations, bananas, oranges, apples, dates, tomatoes, melons, watermelons, figs, olives, grapes, peaches, peafowls, pheasants, multicoloured butterflies, or millions of other living beings such as these. Indeed, they could not obtain even a single cell of any one of them.

Briefly, unconscious atoms cannot form the cell by coming together. They cannot take a new decision and divide this cell into two, then take other decisions and create the professors who first invent the electron microscope and then examine their own cell structure under that microscope. **Matter is an unconscious, lifeless heap, and it comes to life with God's superior creation.**

Evolutionary theory, which claims the opposite, is a total fallacy completely contrary to reason. Thinking even a little bit on the claims of tevolutionists discloses this reality, just as in the above example.

Technology In The Eye and The Ear
Another subject that remains unanswered by evolutionary theory is the excellent quality of perception in the eye and the ear.

Before passing on to the subject of the eye, let us briefly answer the question of "how we see". Light rays coming from an object fall oppositely on the retina of the eye. Here, these light rays are transmitted into electric signals by cells and they reach a tiny spot at the back of the brain called the centre of vision. These electric signals are perceived in this centre of the brain as an image after a series of processes. With this technical background, let us do some thinking.

The brain is insulated from light. That means that the inside of the brain is solid dark, and light does not reach the location where the brain is situated. The place called the centre of vision is a solid dark place where no light ever reaches; it may even be the darkest place you have ever known. However, you observe a luminous, bright world in this pitch darkness.

The image formed in the eye is so sharp and distinct that even the technology of the 20th century has not been able to attain it. For instance, look at the book you read, your hands with which you hold it, then lift your head and look around you. Have you ever seen such a sharp and distinct image as this one at any other place? Even the most developed television screen produced by the greatest television producer in the world cannot provide such a sharp image for you. This is a three-dimensional, coloured, and extremely sharp image. For more than 100 years, thousands of engineers have been trying to achieve this sharpness. Factories, huge premises were established, much research has been done, plans and designs have been made for this purpose. Again, look at a TV screen and the book you hold in your hands. You will see that there is a big difference in sharpness and distinction. Moreover, the TV screen shows you a two-dimensional image, whereas with your eyes, you watch a three-dimensional perspective having depth. When you look carefully, you will see that there is a blurring in the television, is there any blurring in your vision? Surely there is not.

For many years, ten of thousands of engineers have tried to make a three-dimensional TV, and reach the vision quality of the eye. Yes, they have made a three-dimensional television system but it is not possible to watch it without putting on glasses; moreover, it is only an artificial three-dimension. The background is more blurred, the foreground appears like a paper setting. Never has it been possible to produce a sharp and distinct vision like that of the eye. In both the camera and the television, there is a loss of image quality.

Evolutionists claim that the mechanism producing this sharp and distinct image has been formed by chance. Now, if somebody told you that the television in your room was formed as a result of chance, that all its atoms just happened to come together and make up this device that produces an image, what would you think? How can atoms do what thousands of people cannot?

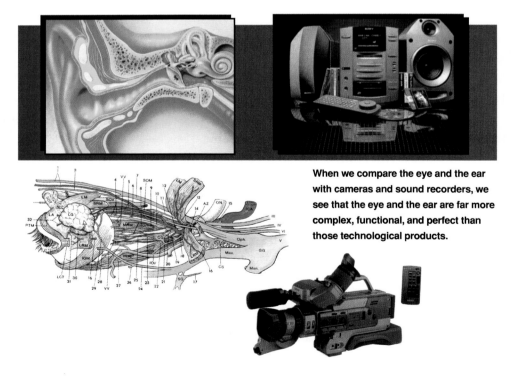

When we compare the eye and the ear with cameras and sound recorders, we see that the eye and the ear are far more complex, functional, and perfect than those technological products.

For nearly a century, tens of thousands of engineers have been researching and striving in high-tech laboratories and great industrial complexes using the most advanced technological devices, and they have been able to do no more than this.

If a device producing a more primitive image than the eye could not have been formed by chance, then it is very evident that the eye and the image seen by the eye could not have been formed by chance. It requires a much more detailed and wise plan and design than the one in the TV. The plan and design of the image as distinct and sharp as this one belongs to God, Who has power over all things.

The same situation applies to the ear. The outer ear picks up the available sounds by the auricle and directs them to the middle ear; the middle ear transmits the sound vibrations by intensifying them; the inner ear sends these vibrations to the brain by translating them into electric signals. Just as with the eye, the act of hearing finalises in the centre of hearing in the brain.

The situation in the eye is also true for the ear. That is, the brain is insulated from sound just like it is from light: it does not let any sound in. Therefore, no matter how noisy is the outside, the inside of the brain is completely silent. Nevertheless, the sharpest sounds are perceived in the brain. In your brain, which is insulated from sound, you listen to the symphonies of an orchestra, and hear all the noises in a crowded place. However, if the sound level in your brain was measured by a precise device at that moment, it would be seen that a complete silence is prevailing there.

Let us again compare the high quality and superior technology present in the ear and the brain with the technology produced by human beings. As is the case with imagery, decades of effort have been spent in trying to generate and reproduce sound that is faithful to the original. The results of these efforts are sound recorders, high-fidelity systems, and systems for sensing sound. Despite all this technology and the thousands of engineers and experts who have been working in this endeavour, no sound has yet been obtained that has the same sharpness and clarity as the sound perceived by the ear. Think of the highest-quality HI-FI systems produced by the biggest company in the music industry. Even in these devices, when sound is recorded some of it is lost; or when you turn on the HI-FI you always hear a hissing sound before the music starts. However, the sounds that are the products of the technology of the human body are extremely sharp and clear. A human ear never perceives a sound accompanied by a hissing sound or with atmospherics as a HI-FI does; it perceives the sound exactly as it is, sharp and clear. This is the way it has been since the creation of man.

Briefly, the technology in our body is far superior to the technology mankind has produced using its accumulated information, experience, and opportunities. No one would say that a HI-FI or a camera came into being as a result of chance. So how can it be claimed that the technologies that exist in the human body, which are superior even to these, could have come into being as a result of a chain of coincidences called evolution?

It is evident that the eye, the ear, and indeed all the other parts of the human body are products of a very superior creation. These are crystal-clear indications of God's unique and unmatched creation, of His eternal knowledge and might.

The reason we specifically mention the senses of seeing and hearing here is the inability of evolutionists to understand evidence of creation so clear as this. If, one day, you ask an evolutionist to explain to you how this excellent design and technology became possible in the eye and the ear as a result of chance, you will see that he will not be able to give you any reasonable or logical reply. Even **Darwin**, in his letter to Asa Gray on April 3rd 1860, wrote that "**the thought of the eye made him cold all over**" and he confessed the desperation of the evolutionists in the face of the excellent design of living things.[149]

The Theory of Evolution is the Most Potent Spell in the World

Throughout this book it has been explained that the theory of evolution lacks any scientific evidence and that on the contrary, scientific proofs from such branches of science such as paleontology, microbiology and anatomy reveal it to be a bankrupt theory. It has been stressed that evolution is incompatible with scientific discoveries, reason and logic.

It needs to be made clear that anyone free of prejudice and the influence of any particular ideology, who uses only his reason and logic, will clearly understand that belief in the theory of evolution, which brings to mind the superstitions of societies with no knowledge of science or civilization, is quite impossible.

As has been explained above, those who believe in the theory of evolution think that a few atoms and molecules thrown into a huge vat could produce thinking, reasoning professors, university students, scientists such as Einstein and Galileo, artists such as Humphrey Bogart, Frank Sinatra and Pavarotti, as well as antelopes, lemon trees and carnations. Moreover, the scientists and professors who believe in this nonsense are educated people. That is why it is quite justifiable to speak of the theory of evolution as "the most potent spell in history." Never before has any other belief or idea so taken away peoples' powers of reason, refused to allow them to think intelligently and logically and hidden the truth from them as if they had been blindfolded. This is an even worse and unbelievable blindness than the Egyptians worshipping the Sun God Ra, totem worship in some parts of Africa, the people of Saba worshipping the Sun, the tribe of the Prophet Abraham worshipping idols they had made with their own hands or the people of the Prophet Moses worshipping the Golden Calf.

In fact, this situation is a lack of reason God points out in the Qur'an. He reveals in many verses that some peoples' minds will be closed and that they will be powerless to see the truth. Some of these verses are as follows:

> **As for those who disbelieve, it makes no difference to them whether you warn them or do not warn them, they will not believe. God has sealed up their hearts and hearing and over their eyes is a blindfold. They will have a terrible punishment. (Surat al-Baqara: 6-7)**

> …They have hearts they do not understand with. They have eyes they do not see with. They have ears they do not hear with. Such people are like cattle. No, they are even further astray! They are the unaware. (Surat al-A'raf: 179)

> Even if We opened up to them a door into heaven, and they spent the day ascending through it, they would only say, "Our eyesight is befuddled! Or rather we have been put under a spell!" (Surat al-Hijr: 14-15)

Words cannot express just how astonishing it is that this spell should hold such a wide community in thrall, keep people from the truth, and not be broken for 150 years. It is understandable that one or a few people might believe in impossible scenarios and claims full of stupidity and illogicality. However, "magic" is the only possible explanation for people from all over the world believing that unconscious and lifeless atoms suddenly decided to come together and form a universe that functions with a flawless system of organization, discipline, reason and consciousness, the planet Earth with all its features so perfectly suited to life, and living things full of countless complex systems.

In fact, God reveals in the Qur'an in the incident of the Prophet Moses and Pharaoh that some people who support atheistic philosophies actually influence others by magic. When Pharaoh was told about the true religion, he told the Prophet Moses to meet with his own magicians. When the Prophet Moses did so, he told them to demonstrate their abilities first. The verses continue:

> He said, "You throw." And when they threw, they cast a spell on the people's eyes and caused them to feel great fear of them. They produced an extremely powerful magic. (Surat al-A'raf: 116)

As we have seen, Pharaoh's magicians were able to deceive everyone, apart from the Prophet Moses and those who believed in him. However, the evidence put forward by the Prophet Moses broke that spell, or "swallowed up what they had forged" as the verse puts it.

> We revealed to Moses, "Throw down your staff." And it immediately swallowed up what they had forged. So the Truth took place and what they did was shown to be false. (Surat al-A'raf: 117-119)

As we can see from that verse, when it was realised that what these people who had first cast a spell over others had done was just an illusion, they lost all credibility. In the present day too, unless those who under the influence of a similar spell believe in these ridiculous claims under their scientific disguise and spend their lives defending them abandon them, they too will be humiliated when the full truth emerges and the spell is broken. In fact, Malcolm Muggeridge, an atheist philosopher and supporter of evolution admitted he was worried by just that prospect:

> I myself am convinced that the theory of evolution, especially the extent to which it's been applied, will be one of the great jokes in the history books in the future. Posterity will marvel that so very flimsy and dubious an hypothesis could be accepted with the incredible credulity that it has.[150]

That future is not far off: On the contrary, people will soon see that "chance" is not a god, and will look back on the theory of evolution as the worst deceit and the most terrible spell in the world. That spell is already rapidly beginning to be lifted from the shoulders of people all over the world. Many people who see the true face of the theory of evolution are wondering with amazement how it was that they were ever taken in by it.

CHAPTER 13

Evolutionist Claims
and the Facts

In previous chapters, we examined the invalidity of the theory of evolution in terms of the bodies of evidence found in fossils and from the standpoint of molecular biology. In this chapter, we will address a number of biological phenomena and concepts presented as theoretical evidence by evolutionists. These topics are particularly important for they show that there is no scientific finding that supports evolution and instead reveal the extent of the distortion and hoodwink employed by evolutionists.

Variations and Species

Variation, a term used in genetics, refers to a genetic event that causes the individuals or groups of a certain type or species to possess different characteristics from one another. For example, all the people on earth carry basically the same genetic information, yet some have slanted eyes, some have red hair, some have long noses, and others are short of stature, all depending on the extent of the variation potential of this genetic information.

Evolutionists predicate the variations within a species as evidence to the theory. However, **variation does not constitute evidence for evolution because variations are but the outcomes of different combinations of already existing genetic information and they do not add any new characteristic to the genetic information**. The important thing for the theory of evolution, however, is the question of how brand-new information to make a brand-new species could come about.

Variation always takes place within the limits of genetic information. In the science of genetics, this limit is called the "gene pool". All of the characteristics present in the gene pool of a species may come to light in various ways due to variation. For example, as a result of variation, varieties that have relatively longer tails or shorter legs may appear in a certain species of reptile, since information for both long-legged and short-legged forms may exist in the gene pool that species. However, variations do not transform reptiles into birds by adding wings or feathers to them, or by changing their metabolism. Such a change requires an increase in the genetic information of the living thing, which is certainly not possible through variations.

Darwin was not aware of this fact when he formulated his theory. He thought that there was no limit to variations. In an article he wrote in 1844 he stated: "**That a limit to variation does exist in nature is assumed by most authors, though I am unable to discover a single fact on which this belief is grounded**".[151] In *The Origin of Species* he cited different examples of variations as the most important evidence for his theory.

For instance, according to Darwin, animal breeders who mated different varieties of cattle in order to bring about new varieties that produced more milk, were ultimately going to transform them into a different species. Darwin's notion of "unlimited variation" is best seen in the following sentence from *The Origin of Species*:

> I can see no difficulty in a race of bears being rendered, by natural selection, more and more aquatic in their habits, with larger and larger mouths, till a creature was produced as monstrous as a whale.[152]

The reason Darwin cited such a far-fetched example was the primitive understanding of science in his day. Since then, in the 20th century, science has posited the principle of "**genetic stability**" (genetic homeostasis), based on the results of experiments conducted on living things. This principle holds that, since all mating attempts carried out to produce new variations have been inconclusive, there are **strict barriers among different species of living things**. This meant that it was absolutely impossible for animal breeders to convert cattle into a different species by mating different variations of them, as Darwin had postulated.

Norman Macbeth, who disproved Darwinism in his book *Darwin Retried*, states:

> The heart of the problem is whether living things do indeed vary to an unlimited extent... **The species look stable.** We have all heard of disappointed breeders who carried their work to a certain point only to see the animals or

DID WHALES EVOLVE FROM BEARS?
In The Origin of Species, Darwin asserted that whales had evolved from bears that tried to swim! Darwin mistakenly supposed that the possibilities of variation within a species were unlimited. 20th century science has shown this evolutionary scenario to be imaginary.

plants revert to where they had started. Despite strenuous efforts for two or three centuries, it has never been possible to produce a blue rose or a black tulip.[153]

Luther Burbank, considered the most competent breeder of all time, expressed this fact when he said, "there are limits to the development possible, and these limits follow a law." [154] The Danish scientist W. L. Johannsen sums the matter up this way:

> The variations upon which Darwin and Wallace had placed their emphasis cannot be selectively pushed beyond a certain point, that **such a variability does not contain the secret of 'indefinite departure**.[155]

Antibiotic Resistance and DDT Immunity are not Evidence for Evolution

One of the biological concepts that evolutionists try to present as evidence for their theory is the resistance of bacteria to antibiotics. Many evolutionist sources show antibiotic resistance as "an example of the development of living things by advantageous mutations". A similar claim is also made for the insects which build immunity to insecticides such as DDT.

However, evolutionists are mistaken on this subject too.

Antibiotics are "killer molecules" that are produced by micro-organisms to fight other micro-organisms. The first antibiotic was penicillin, discovered by Alexander Fleming in 1928. Fleming realized that mould produced a molecule that killed the *Staphylococcus* bacterium, and this discovery marked a turning point in the world of medicine. Antibiotics derived from micro-organisms were used against bacteria and the results were successful.

Soon, something new was discovered. Bacteria build immunity to antibiotics over time. The mechanism works like this: A large proportion of the bacteria that are subjected to antibiotics die, but some others, which are not affected by that antibiotic, replicate rapidly and soon make up the whole population. Thus, the entire population becomes immune to antibiotics.

Evolutionists try to present this as "the evolution of bacteria by adapting to conditions".

The truth, however, is very different from this superficial interpretation. One of the scientists who has done the most detailed research into this subject is the Israeli biophysicist Lee Spetner, who is also known for his book *Not by Chance* published in 1997. Spetner maintains that the immunity of bacteria comes about by two different mechanisms, but neither of them constitutes evidence for the theory of evolution. These two mechanisms are:

1) The transfer of resistance genes already extant in bacteria.

2) The building of resistance as a result of losing genetic data because of mutation.

Professor Spetner explains the first mechanism in an article published in 2001:

> Some microorganisms are endowed with genes that grant resistance to these antibiotics. This resistance can take the form of degrading the antibiotic molecule or of ejecting it from the cell... The organisms having these genes can transfer them to other bacteria making them resistant as well. Although the resistance mechanisms are specific to a particular antibiotic, most pathogenic bacteria have... succeeded in accumulating several sets of genes granting them resistance to a variety of antibiotics.[156]

Spetner then goes on to say that this is not "evidence for evolution":

> The acquisition of antibiotic resistance in this manner... is not the kind that can serve as a prototype for the mutations needed to account for Evolution. The genetic changes that could illustrate the theory must not only add information to the bacterium's genome, they must add new information to the biocosm. The horizontal transfer of genes only spreads around genes that are already in some species.[157]

So, we cannot talk of any evolution here, because no new genetic information is produced: genetic information that already exists is simply transferred between bacteria.

The second type of immunity, which comes about as a result of mutation, is not an example of evolution either. Spetner writes:

> ...A microorganism can sometimes acquire resistance to an antibiotic through a random substitution of a single nucleotide... Streptomycin, which was discovered by Selman Waksman and Albert Schatz and first reported in 1944, is an antibiotic against which bacteria can acquire resistance in this way. But although the mutation they undergo in the process is beneficial to the microorganism in the presence of streptomycin, it cannot serve as a prototype for the kind of mutations needed by NDT [Neo Darwinian Theory]. The type of mutation that grants resistance to streptomycin is manifest in the ribosome and degrades its molecular match with the antibiotic molecule. This change in the surface of the microorganism's ribosome prevents the streptomycin molecule from attaching and carrying out its antibiotic function. It turns out that this degradation is a loss of specificity and therefore a loss of information. The main point is that (Evolution) cannot be achieved by mutations of this sort, no matter how many of them there are. Evolution cannot be built by accumulating mutations that only degrade specificity.[158]

To sum up, a mutation impinging on a bacterium's ribosome makes that bacterium resistant to streptomycin. The reason for this is the "decomposition" of the ribosome by mutation. That is, no new genetic information is added to the bacterium. On the contrary, the structure of the ribosome is decomposed, that is to say, the bacterium becomes "disabled". (Also, it has been discovered that the ribosome of the mutated bacterium is less functional than that of normal bacterium). Since this "disability" prevents the antibiotic from attaching onto the ribosome, "antibiotic resistance" develops.

Finally, there is no example of mutation that "develops the genetic information".

The same situation holds true for the immunity that insects develop to DDT and similar insecticides. In most of these instances, immunity genes that already exist are used. The evolutionist biologist Francisco Ayala admits this fact, saying, "The genetic variants required for resistance to the most diverse kinds of pesticides were apparently present in every one of the populations exposed to these man-made compounds."[159] Some other examples explained by mutation, just as with the ribosome mutation mentioned above, are phenomena that cause "genetic information deficit" in insects.

In this case, it cannot be claimed that the immunity mechanisms in bacteria and insects constitute evidence for the theory of evolution. That is because the theory of evolution is based on the assertion that living things develop through mutations. However, Spetner explains that neither antibiotic immunity nor any other biological phenomena indicate such an example of mutation:

> The mutations needed for macroevolution have never been observed. No random mutations that could represent the mutations required by Neo-Darwinian Theory that have been examined on the molecular level have added any information. The question I address is: Are the mutations that have been observed the kind the theory needs for support? The answer turns out to be NO![160]

The Fallacy of Vestigial Organs

For a long time, the concept of "vestigial organs" appeared frequently in evolutionist literature as "evidence" of evolution. Eventually, it was silently put to rest when this was proved to be invalid. But some evolutionists still believe in it, and from time to time someone will try to advance "vestigial organs" as important evidence of evolution.

The notion of "vestigial organs" was first put forward a century ago. As evolutionists would have it, there existed in the bodies of some creatures a number of non-functional organs. These had been inherited from progenitors and had gradually become vestigial from lack of use.

The whole assumption is quite unscientific, and is based entirely on insufficient knowledge. These "**non-functional organs**" were in fact organs whose "**functions had not yet been discovered**". The best indication of this was the gradual yet substantial decrease in evolutionists' long list of vestigial organs. S.R. Scadding, an evolutionist himself, concurred with this fact in his article "Can vestigial organs constitute evidence for evolution?" published in the journal *Evolutionary Theory*:

> Since it is not possible to unambiguously identify useless structures, and since the structure of the argument used is not scientifically valid, **I conclude that "vestigial organs" provide no special evidence for the theory of evolution.**[161]

The list of vestigial organs that was made by the German Anatomist R. Wiedersheim in 1895 included approximately 100 organs, including the appendix and coccyx. As science progressed, it was discovered that all of the organs in Wiedersheim's list in fact had very important functions. For instance, it was discovered that the appendix, which was supposed to be a "vestigial organ", was in fact a lymphoid organ that fought against infections in the body. This fact was made clear in 1997: "Other bodily organs and tissues-the thymus, liver, spleen, **appendix**, bone marrow, and small collections of lymphatic tissue such as the tonsils in the throat and Peyer's patch in the small intestine-are also part of the lymphatic system. They too **help the body fight infection.**"[162]

It was also discovered that the **tonsils**, which were included in the same list of vestigial organs, had a significant role in protecting the throat against infections, particularly until adolescence. It was found that the **coccyx** at the lower end of the vertebral column supports the bones around the pelvis and is the convergence point of some small muscles and for this reason, it would not be possible to sit comfortably without a coccyx. In the years that followed, it was realised that the **thymus** triggered the immune system in the human body by activating the T cells, that the **pineal gland** was in charge of the secretion of some important hormones, that the **thyroid gland** was effective in providing steady growth in babies and children, and that the **pituitary gland** controlled the correct functioning of many hormone glands. All of these were once conside-red to be "vestigial organs".

All instances of vestigial organs have been disproved in time. For example the semicircular fold in the eye, which was mentioned in the *Origins* as a vestigial structure, has been shown to be fully functional in our time, though its function was unknown in Darwin's time. This organ lubricates the eyeball.

Finally, the semi-lunar fold in the eye, which was referred to as a vestigial organ by Darwin, has been found in fact to be in charge of cleansing and lubricating the eyeball.

There was a very important logical error in the evolutionist claim regarding vestigial organs. As we have just seen, this claim was that the vestigial organs in living things were inherited from their ancestors. However, some of the alleged "vestigial" organs are not found in the species alleged to be the ancestors of human beings! For example, the appendix does not exist in some ape species that are said to be ancestors of man. The famous biologist H. Enoch, who challenged the theory of vestigial organs, expressed this logical error as follows:

> Apes possess an appendix, whereas their less immediate relatives, the lower apes, do not; but it appears again among the still lower mammals such as the opossum. How can the evolutionists account for this?[163]

Simply put, the scenario of vestigial organs put forward by evolutionists contains a number of serious logical flaws, and has in any case been proven to be scientifically untrue. There exists not one inherited vestigial organ in the human body, since human beings did not evolve from other creatures as a result of chance, but were created in their current, complete, and perfect form.

The Myth of Homology

Structural similarities between different species are called "**homology**" in biology. Evolutionists try to present those similarities as evidence for evolution.

Darwin thought that creatures with similar (homologous) organs had an evolutionary relationship with each other, and that these organs must have been inherited from a common ancestor. According to his assumption, both pigeons and eagles had wings; therefore, pigeons, eagles, and indeed all other birds with wings were supposed to have evolved from a common ancestor.

Homology is a deceptive argument, advanced on the basis of no other evidence than an apparent physical resemblance. This argument has never once been verified by a single concrete discovery in all the years since Darwin's day. Nowhere in the world has anyone come up with a fossil remain of the imaginary common ancestor of creatures with homologous structures. Furthermore, the following issues make it clear that homology provides no evidence that evolution ever occurred.

1. One finds homologous organs in creatures belonging to completely different phyla, among which evolutionists have not been able to establish any sort of evolutionary relationship;

2. The genetic codes of some creatures that have homologous organs are completely different from one another.

3. The embryological development of homologous organs in different creatures is completely different.

Let us now examine each of these points one by one.

Similar Organs in Entirely Different Living Species

There are a number of homologous organs shared by different groups among which evolutionists cannot establish any kind of evolutionary relationship. Wings are one example. In addition to birds, we find wings on bats, which are mammals, and on insects and even on some dinosaurs, which are extinct reptiles. Not even evolutionists posit an evolutionary relationship or kinship among those four different groups of animals.

Another striking example is the amazing resemblance and the structural similarity observed in the eyes of different creatures. For example, the octopus and man are two extremely different species, between which no evolutionary relationship is likely even to be proposed, yet the eyes of both are very much alike in terms of their structure and function. Not even evolutionists try to account for the similarity of the eyes of the octopus and man by positing a common ancestor. These and numerous other examples show that the evolutionist claim based on resemblances is completely unscientific.

In fact, homologous organs should be a great embarrassment for evolutionists. The famous evolutionist Frank Salisbury's confessions revealed in his statements on how extremely different creatures came to have very similar eyes underscores the impasse of homology:

> Even something as complex as **the eye has appeared several times**; for example, in the squid, the vertebrates, and the arthropods. It's bad enough accounting for the origin of such things once, but the **thought of producing them several times according to the modern synthetic theory makes my head swim.**[164]

There are many creatures which, despite their very similar physical make-up, do not permit any claims of evolutionary relationship. Two large mammal categories, placentals and marsupials, are an example. Evolutionists consider this distinction to have come about when mammals first appeared, and that each group lived its own evolutionary history totally independent of the other. But it is interesting that there are "pairs" in placentals and marsupials which are nearly the same. The American biologists Dean Kenyon and Percival Davis make the following comment:

> According to Darwinian theory, the pattern for wolves, cats, squirrels, ground hogs, anteaters, moles, and mice each evolved twice: once in placental mammals and again, totally independently, in marsupials. This amounts to the astonishing claim that a random, undirected process of mutation and natural selection somehow hit upon identical features several times in widely separated organisms.[165]

Extraordinary resemblances and similar organs like these, which evolutionist biologists cannot accept as examples of "homology," show that there is no evidence for the thesis of evolution from a common ancestor.

The Genetic and Embryological Impasse of Homology

In order for the evolutionist claim concerning "homology" to be taken seriously, similar (homologous) organs in different creatures should also be coded with similar (homologous) DNA codes. However, they are not. Similar organs are usually governed by very different genetic (DNA) codes. Furthermore, similar genetic codes in the DNA of different creatures are often associated with completely different organs.

Michael Denton, an Australian professor of biochemistry, describes in his book *Evolution: A Theory in Crisis* the genetic impasse of the evolutionist interpretation of homology: "**Homologous structures are often specified by non-homologous genetic systems** and the concept of homology can seldom be extended back into embryology."[166]

A famous example on this subject is the "five digit skeletal structure" of quadrupeds which is quoted in almost all evolutionist textbooks. Quadrupeds, i.e., land-living vertebrates, have five digits on their fore- and hindlimbs. Although these do not always have the appearance of five digits as we know them, they are all counted as pentadactyl due to their bone structure. The fore- and hindlimbs of a frog, a lizard, a squirrel or a monkey all have this same structure. Even the bone structures of birds and bats conform to this basic design.

Evolutionists claim that all living things descended from a common ancestor, and they have long cited pentadactyl limb as evidence of this.

This claim was mentioned in almost all basic sources on biology throughout the 20th century as very strong evidence for evolution. Genetic findings in the 1980s refuted this evolutionist claim. It was realised that the pentadactyl limb patterns of different creatures are controlled by totally different genes. Evolutionist biologist William Fix describes the collapse of the evolutionist thesis regarding pentadactylism in this way:

> The older text-books on evolution make much of the idea of homology, pointing out the obvious resemblances between the skeletons of the limbs of different animals. Thus the "pentadactyl" limb pattern is found in the arm of a man, the wing of a bird, and the flipper of a whale, and this is held to indicate their common origin. Now if these various structures were transmitted by the same gene couples, varied from time to time by mutations and acted upon by environmental selection, the theory would make good sense. Unfortunately this is not the case. Homologous organs are now known to be produced by totally different gene complexes in the different species. The concept of homology in terms of similar genes handed on from a common ancestor has broken down...[167]

Another point is that in order for the evolutionary thesis regarding homology to be taken seriously, the periods of similar structures' embryological development-in other words, the stages of development in the egg or the mother's womb-would need to be parallel, whereas, in reality, these embryological periods for similar structures are quite different from each other in every living creature.

To conclude, we can say that genetic and embryological research has proven that the concept of homology defined by Darwin as "evidence of the evolution of living things from a common ancestor" can by no means be regarded as any evidence at all. In this respect, science can be said to have proven the Darwinist thesis false time and time again.

Invalidity of the Claim of Molecular Homology

Evolutionists' advancement of homology as evidence for evolution is invalid not only at the morphological level, but also at the molecular level. Evolutionists say that the **DNA codes, or the corresponding protein structures,** of different living species are similar, and that this similarity is evidence that these living species have evolved from common ancestors, or else from each other.

In truth, however, the results of molecular comparisons do not work in favour of the theory of evolution at all. There are huge molecular differences between creatures that appear to be very similar and related. For instance, the cytochrome-C protein, one of the proteins vital to respiration, is incredibly different in living beings of the same class. According to research carried out on this matter, the difference between two different reptile species is greater than the difference between a bird and a fish or a fish and a mammal. Another study has shown that molecular differences between some birds are greater than the differences between those same birds and mammals. It has also been discovered that the molecular difference between bacteria that appear to be very similar is greater than the difference between mammals and amphibians or insects.[168] Similar comparisons have been made in the cases of haemoglobin, myoglobin, hormones, and genes and similar conclusions are drawn.[169]

Concerning these findings in the field of molecular biology, Dr. Michael Denton comments:

> **Each class at a molecular level is unique, isolated and unlinked by intermediates**. Thus, molecules, like fossils, have failed to provide the elusive intermediates so long sought by evolutionary biology... **At a molecular level, no organism is "ancestral" or "primitive" or "advanced" compared with its relatives...** There is little doubt that if this molecular evidence had been available a century ago... the idea of organic evolution might never have been accepted.[170]

Professor Michael Denton: "Evolution is a theory in crisis"

In the 1990s, research into the genetic codes of living things worsened the quandary faced by the theory of evolution in this regard. In these experiments, instead of the earlier comparisons that were limited to protein sequences, "ribosomal RNA" (rRNA) sequences were compared. From these findings, evolutionist scientists sought to establish an "evolutionary tree". However, they were disappointed by the results. According to a 1999 article by French biologists Hervé Philippe and Patrick Forterre, "with more and more sequences available, it turned out that **most protein pyhlogenies contradict each other as well as the rRNA tree**."[171]

Besides rRNA comparisons, the DNA codes in the genes of living things were also compared, but the results have been the opposite of the "tree of life" presupposed by evolution. Molecular biologists James A. Lake, Ravi Jain and Maria C. Rivera elaborated on this in an article in 1999:

> "Scientists started analyzing a variety of genes from different organisms and

found that their relationship to each other contradicted the evolutionary tree of life derived from rRNA analysis alone."172

Neither the comparisons that have been made of proteins, nor those of rRNAs or of genes, confirm the premises of the theory of evolution. Carl Woese, a highly reputed biologist from the University of Illinois admits that the concept of "phylogeny" has lost its meaning in the face of molecular findings in this way:

> **No consistent organismal phylogeny has emerged from the many individual protein phylogenies so far produced**. Phylogenetic incongruities can be seen everywhere in the universal tree, from its root to the major branchings within and among the various (groups) to the makeup of the primary groupings themselves."173

The fact that results of molecular comparisons are not in favour of, but rather opposed to, the theory of evolution is also admitted in an article called "Is it Time to Uproot the Tree of Life?" published in *Science* in 1999. This article by Elizabeth Pennisi states that the genetic analyses and comparisons carried out by Darwinist biologists in order to shed light on the "tree of life" actually yielded directly opposite results, and goes on to say that "new data are muddying the evolutionary picture":

> A year ago, biologists looking over newly sequenced genomes from more than a dozen microorganisms thought these data might support the accepted plot lines of life's early history. But what they saw confounded them. Comparisons of the genomes then available not only didn't clarify the picture of how life's major groupings evolved, they confused it. And now, with an additional eight microbial sequences in hand, the situation has gotten even more confusing.... Many evolutionary biologists had thought they could roughly see the beginnings of life's three kingdoms... When full DNA sequences opened the way to comparing other kinds of genes, researchers expected that they would simply add detail to this tree. But "nothing could be further from the truth," says Claire Fraser, head of The Institute for Genomic Research (TIGR) in Rockville, Maryland. Instead, the **comparisons have yielded many versions of the tree of life that differ from the rRNA tree and conflict with each other as well**...174

In short, as molecular biology advances, the homology concept loses more ground. Comparisons that have been made of proteins, rRNAs and genes reveal that creatures which are allegedly close relatives according to the theory of evolution are actually totally distinct from each other. A 1996 study using 88 protein sequences grouped rabbits with primates instead of rodents; a 1998 analysis of 13 genes in 19 animal species placed sea urchins among the chordates; and another 1998 study based on 12 proteins put cows closer to whales than to horses. Molecular biologist Jonathan Wells sums up the situation in 2000 in this way:

> Inconsistencies among trees based on different molecules, and the bizarre trees that result from some molecular analyses, have now plunged molecular phylogeny into a crisis.[175]

The Myth of Embryological Recapitulation

What used to be called the "recapitulation theory" has long been eliminated from scientific literature, but it is still being presented as a scientific reality by some evolutionist publications. The term "recapitulation" is a condensation of the dictum "Ontogeny recapitulates phylogeny", put forward by the evolutionist biologist Ernst Haeckel at the end of the 19th century.

This theory of Haeckel's postulates that living embryos re-experience the evolutionary process that their pseudo-ancestors underwent. He theorised that during its development in its mother's womb, the human embryo first displayed the characteristics of a fish, and then those of a reptile, and finally those of a human.

It has since **been proven that this theory is completely bogus**. It is now known that the "gills" that supposedly appear in the early stages of the human embryo are in fact the initial phases of the middle-ear canal, parathyroid, and thymus. The part of the embryo that was likened to the "egg yolk pouch" turns out to be a pouch that produces blood for the infant. The part that had been identified as a "tail" by Haeckel and his followers is in fact the backbone, which resembles a tail only because it takes shape before the legs do.

Haeckel was an evolutionist even more ardent than Darwin in many respects. For this reason, he did not hesitate to distort the scientific data and devise various forgeries.

These are universally acknowledged facts in the scientific world, and are accepted even by evolutionists themselves. George Gaylord Simpson, one of the founders of neo-Darwinism, writes:

HUMAN EMBRYOS DO NOT HAVE GILL SLITS

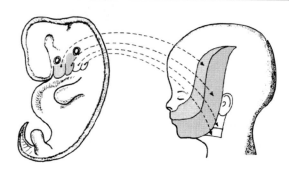

Once defined as an inheritance from past ancestors, the folds on the human embryos are now redefined. It has been shown that human embryos do not recapitulate evolutionary history of man.

Haeckel misstated the evolutionary principle involved. **It is now firmly established that ontogeny does not repeat phylogeny.**[176]

In an article published in *American Scientist*, we read:

Surely **the biogenetic law is as dead as a doornail**. It was finally exorcised from biology textbooks in the fifties. As a topic of serious theoretical inquiry it was extinct in the twenties...[177]

Another interesting aspect of "recapitulation" was Ernst Haeckel himself, a faker who falsified his drawings in order to support the theory he advanced. **Haeckel**'s forgeries purported to show that fish and human embryos resembled one another. When he was caught out, the only defence he offered was that other evolutionists had committed similar offences:

> **After this compromising confession of "forgery" I should be obliged to consider myself condemned and annihilated** if I had not the consolation of seeing side by side with me in the prisoners' dock hundreds of fellow culprits, among them many of the most trusted observers and most esteemed biologists. The great majority of all the diagrams in the best biological textbooks, treatises and journals would incur in the same degree the charge of "forgery", for all of them are inexact, and are more or less doctored, schematised and constructed.[178]

There are indeed "hundreds of fellow culprits, among them many of the most trusted observers and most esteemed biologists" whose studies are full of prejudiced conclusions, distortions, and even forgeries. This is because they have all conditioned themselves to champion evolutionary theory although there is not a shred of scientific evidence supporting it.

CHAPTER 14

The Theory of Evolution: A Materialistic Liability

The information we have considered throughout this book has shown us that the theory of evolution has no scientific basis, and that, on the contrary, evolutionist claims conflict with scientific facts. In other words, the force that keeps evolution alive is not science. Evolution may be maintained by some "scientists", but behind it there is another influence at work.

This other influence is materialist philosophy.

Materialist philosophy is one of the oldest beliefs in the world, and assumes the existence of matter as its basic principle. According to this view, matter has always existed, and everything that exists consists of matter. This makes belief in a Creator impossible, of course, because if matter has always existed, and if everything consists of matter, then there can be no supramaterial Creator who created it. Materialism has therefore long been hostile to religious beliefs of every kind that have faith in God.

So the question becomes one of whether the materialist point of view is correct. One method of testing whether a philosophy is true or false is to investigate the claims it makes about science by using scientific methods. For instance, a philosopher in the 10th century could have claimed that there was a divine tree on the surface of the moon and that all living things actually grew on the branches of this huge tree like fruit, and then fell off onto the earth. Some people might have found this philosophy attractive and believed in it. But in the 20th century, at a time when man has managed to walk on the moon, it is no longer possible to seriously hold such a belief. Whether such a tree exists there or not can be determined by scientific methods, that is, by observation and experiment.

We can therefore investigate by means of scientific methods the materialist claim: that matter has existed for all eternity and that this matter can organise itself without a supramaterial Creator and cause life to begin. When we do this, we see that materialism has already collapsed, because the idea that matter has existed since beginning of time **has been overthrown by the Big Bang theory which shows that the universe was created from nothingness.** The claim that matter organised itself and created life is the claim that we call "the theory of evolution" -which this book has been examining-and which has been shown to have collapsed.

However, if someone is determined to believe in materialism and puts his devotion to materialist philosophy before everything else, then he will act differently. If he is a materialist first and a scientist second, he will not abandon materialism when he sees that evolution is disproved by science. On the contrary, he will attempt to uphold and defend materialism by trying to support evolution, no matter what. This is exactly the predicament that evolutionists defending the theory of evolution find themselves in today.

Interestingly enough, they also confess this fact from time to time. A well-known geneticist and outspoken evolutionist, Richard C. Lewontin from Harvard University, confesses that he is "a materialist first and a scientist second" in these words:

> It is not that the methods and institutions of science somehow compel us accept a material explanation of the phenomenal world, but, on the contrary, that **we are forced by our a priori adherence to material causes** to create an apparatus of investigation and a set of concepts that produce material explanations, no matter how counter-intuitive, no matter how mystifying to the uninitiated. **Moreover, that materialism is absolute, so we cannot allow a Divine Foot in the door.**[179]

The term "a priori" that Lewontin uses here is quite important. This philosophical term refers to a presupposition not based on any experimental knowledge. A thought is "a priori" when you consider it to be correct and accept it as so even if there is no information available to confirm it. As the evolutionist Lewontin frankly states, materialism is an "a priori" commitment for evolutionists, who then try to adapt science to this preconception. Since materialism definitely necessitates denying the existence of a Creator, they embrace the only alternative they have in hand, which is the theory of evolution. It does not matter to such scientists that evolution has been belied by scientific facts, because they have accepted it "a priori" as true.

This prejudiced behaviour leads evolutionists to a belief that "unconscious matter composed itself", which is contrary not only to science, but also to reason. Professor of chemistry from New York University and a DNA expert Robert Shapiro, as we have quoted before, explains this belief of evolutionists and the materialist dogma lying at its base as follows:

> Another evolutionary principle is therefore needed to take us across the gap from mixtures of simple natural chemicals to the first effective replicator. This principle has not yet been described in detail or demonstrated, but it is anticipated, and given names such as chemical evolution and **self-organization of matter. The existence of the principle is taken for granted in the philosophy of dialectical materialism**, as applied to the origin of life by Alexander Oparin.[180]

Evolutionist propaganda, which we constantly come across in the Western media and in well-known and "esteemed" science magazines, is the outcome of this ideological necessity. Since evolution is considered to be indispensable, it has been turned into a sacred cow by the circles that set the standards of science.

Some scientists find themselves in a position where they are forced to defend this far-fetched theory, or at least avoid uttering any word against it, in order to maintain their reputations. Academics in the Western countries have to have articles published in certain scientific journals to attain and hold onto their professorships. All of the journals dealing with biology are under the control of evolutionists, and they do not allow any anti-evolutionist article to appear in them. Biologists, therefore, have to conduct their research under the domination of this theory. They, too, are part of the established order, which regards evolution as an ideological necessity, which is why they blindly defend all the "impossible coincidences" we have been examining in this book.

Materialist Confessions

The German biologist Hoimar von Ditfurth, a prominent evolutionist, is a good example of this bigoted materialist understanding. After Ditfurth cites an example of the extremely complex composition of life, this is what he says concerning the question of whether it could have emerged by chance or not:

> Is such a harmony that emerged only out of coincidences possible in reality? This is the basic question of the whole of biological evolution. Answering this question as "Yes, it is possible" is something like verifying faith in the modern science of nature. Critically speaking, we can say that somebody who accepts

DARWINISM AND MATERIALISM

The only reason that Darwin's theory is still defended despite its obvious refutation by science is the close link between that theory and materialism. Darwin applied materialist philosophy to the natural sciences and the advocates of this philosophy, Marxists being foremost among them, go on defending Darwinism no matter what.

One of the most famous contemporary champions of the theory of evolution, the biologist Douglas Futuyma, wrote: "Together with Marx's materialistic theory of history... Darwin's theory of evolution was a crucial plank in the platform of mechanism and materialism." This is a very clear admission of why the theory of evolution is really so important to its defenders.[1]

Another famous evolutionist, the paleontologist Stephen J. Gould said: "Darwin applied a consistent philosophy of materialism to his interpretation of nature".[2] Leon Trotsky, one of the masterminds of the Russian Communist Revolution along with Lenin, commented: "The discovery by Darwin was the highest triumph of the dialectic in the whole field of organic matter."[3] However, science has shown that Darwinism was not a victory for materialism but rather a sign of that philosophy's overthrow.

1- Douglas Futuyma, *Evolutionary Biology*, 2nd ed., Sunderland, MA: Sinauer, 1986, p. 3
2- Alan Woods and Ted Grant, "Marxism and Darwinism", *Reason in Revolt: Marxism and Modern Science*, London, 1993
3- Alan Woods and Ted Grant. "Marxism and Darwinism", London, 1993

Trotsky

Darwin

Marx

the modern science of nature has no other alternative than to say "yes", because he aims to explain natural phenomena by means that are understandable and tries to derive them from the laws of nature without reverting to supernatural interference. However, at this point, explaining everything by means of the laws of nature, that is, by coincidences, is a sign that he has nowhere else to turn. Because what else could he do other than believe in coincidences?[181]

Yes, as Ditfurth states, the materialist scientific approach adopts as its basic principle explaining life by denying "supernatural interference", i.e. creation. Once this principle is adopted, even the most impossible scenarios are easily accepted. It is possible to find examples of this dogmatic mentality in almost all evolutionist literature. Professor Ali Demirsoy, the well-known advocate of evolutionary theory in Turkey, is just one of many. As we have already pointed out, according to Demirsoy: the probability of the coincidental formation of cythochrome-C, an essential protein for life, is **"as unlikely as the possibility of a monkey writing the history of humanity on a typewriter without making any mistakes"**.[182]

There is no doubt that to accept such a possibility is actually to reject the basic principles of reason and common sense. Even one single correctly formed letter written on a page makes it certain that it was written by a person. When one sees a book of world history, it becomes even more certain that the book has been written by an author. No logical person would agree that the letters in such a huge book could have been put together "by chance".

However, it is very interesting to see that the "evolutionist scientist" Professor Ali Demirsoy accepts this sort of irrational proposition:

> In essence, the probability of the formation of a cytochrome-C sequence is as likely as zero. That is, if life requires a certain sequence, it can be said that this has a probability likely to be realised once in the whole universe. Otherwise some **metaphysical powers** beyond our definition must have acted in its formation. **To accept the latter is not appropriate for the scientific goal**. We thus have to look into the first hypothesis.[183]

Demirsoy writes that he prefers the impossible, in order **"not to have to accept supernatural forces"**-in other words, the existence of a Creator. It is clear that this approach has no relation whatsoever with science. Not surprisingly, when Demirsoy cites another subject-the origins of the mitochondria in the cell-he openly accepts coincidence as an explanation, even though it is "quite contrary to scientific thought".

> The heart of the problem is how the mitochondria have acquired this feature, because attaining this feature by chance even by one individual, requires extreme probabilities that are incomprehensible... The enzymes providing respiration and functioning as a catalyst in each step in a different form make up the core of the mechanism. A cell has to contain this enzyme sequence completely, otherwise it is meaningless. Here, **despite being contrary to biological thought**, in order to avoid a more dogmatic explanation or speculation, we have to accept, though reluctantly, that all the respiration enzymes **completely existed in the cell** before the cell first came in contact with oxygen.[184]

The conclusion to be drawn from such pronouncements is that evolution is not a theory arrived at through scientific investigation. On the con-

The Death of Materialism

Constituting as it does the philosophical underpinnings of the theory of evolution, 19th-century materialism suggested that the universe existed since eternity, that it was not created, and that the organic world could be explained in terms of the interactions of matter. The discoveries of 20th-century science however have completely invalidated these hypotheses.

The supposition that the universe has existed since eternity was blown away by the discovery that the universe originated from a great explosion (the so-called "Big Bang") that took place nearly 15 billion years ago. The Big Bang shows that all physical substances in the universe came into being out of nothing: in other words, they were created. One of the foremost advocates of materialism, the atheist philosopher Anthony Flew concedes:

> Notoriously, confession is good for the soul. I will therefore begin by confessing that the Stratonician atheist has to be embarrassed by the contemporary cosmological consensus (Big Bang). For it seems that the cosmologists are providing a scientific proof ... that the universe had a beginning.[1]

The Big Bang also shows that at each stage, the universe was shaped by a controlled creation. This is made clear by the order that came about after the Big Bang, which was too perfect to have been formed from an uncontrolled explosion. The famous physician Paul Davies explains this situation:

> It is hard to resist the impression that the present structure of the universe, apparently so sensitive to minor alterations in the numbers, has been rather carefully thought out... The seeming miraculous concurrence of numerical values that nature has assigned to her fundamental constants must remain the most compelling evidence for an element of cosmic design.[2]

The same reality makes an American professor of astronomy, George Greenstein, say:

> As we survey all the evidence, the thought insistently arises that some supernatural agency -or rather Agency- must be involved.[3]

Thus, the materialistic hypothesis that life can be explained solely in terms of the interactions of matter also collapsed in the face of the the discoveries of science. In particular, the origin of the genetic information that determines all living things can by no means be explained by any purely material agent. One of the leading defenders of the theory of evolution, George C. Williams, admits this fact in an article he wrote in 1995:

> Evolutionist biologists have failed to realize that they work with two more or less incommensurable domains: that of information and that of matter... the gene is a package of information, not an object... This dearth descriptors makes matter and information two separate domains of existence, which have to be discussed separately, in their own terms.[4]

This situation is evidence for the existence of a supra-material Wisdom that makes genetic information exist. It is impossible for matter to produce information within itself. The director of the German Federal Institute of Physics and Technology, Proffessor Werner Gitt, remarks:

> All experiences indicate that a thinking being voluntarily exercising his own free will, cognition, and creativity, is required. There is no known law of nature, no known process and no known sequence of events which can cause information to originate by itself in matter.[5]

All these scientific facts illustrate that a Creator Who has external power and knowledge, that is, God, creates the universe and all living things. As for materialism, Arthur Koestler, one of the most renowned philosophers of our century says: "It can no longer claim to be a scientific philosophy"[6]

1- Henry Margenau, Roy A. Vargesse. *Cosmos, Bios, Theos.* La Salle IL: Open Court Publishing, 1992, p. 241
2- Paul Davies. *God and the New Physics.* New York: Simon & Schuster, 1983, p. 189
3- Hugh Ross. *The Creator and the Cosmos.* Colorado Springs, CO: Nav-Press, 1993, pp. 114-15
4- George C. Williams. *The Third Culture: Beyond the Scientific Revolution,* New York, Simon & Schuster, 1995, pp. 42-43
5- Werner Gitt. *In the Beginning Was Information.* CLV, Bielefeld, Germany, p. 107, 141
6- Arthur Koestler, *Janus: A Summing Up,* New York, Vintage Books, 1978, p. 250

trary, the form and substance of this theory were dictated by the requirements of materialistic philosophy. It then turned into a belief or dogma in spite of concrete scientific facts. Again, we can clearly see from evolutionist literature that all of this effort has a "purpose"-and that purpose precludes any belief that all living things were not created no matter what the price.

Evolutionists define this purpose as "scientific". However, what they refer to is not science but materialist philosophy. Materialism absolutely rejects the existence of anything "beyond" matter (or of anything supernatural). Science itself is not obliged to accept such a dogma. Science means exploring nature and deriving conclusions from one's findings. If these findings lead to the conclusion that nature is created, science has to accept it. That is the duty of a true scientist; not defending impossible scenarios by clinging to the outdated materialist dogmas of the 19th century.

Materialists, False Religion and True Religion

So far, we have examined how the circles devoted to materialist philosophy derange science, how they deceive people for the sake of the evolutionist fables that they blindly believe, and how they veil realities. That said, we also have to admit that these materialist circles perform a significant "service", though unintentionally.

They carry out this "service", by which they seek to justify their own untrue and atheist thoughts, by exposing all the senselessness and inconsistencies of the traditionalist and bigoted thought that poses in the name of Islam. The offences of the materialist-atheist circle have helped reveal the false religion which has no relation whatsoever with the Qur'an or Islam; which depends on hearsay, superstition, and idle talk; and which has no consistent argument to put forth. Thus, all the inconsistencies, discrepancies, and illogic of the false religion defended by those insincere circles that wrongly act in the name of Islam without relying on valid evidence are exposed.

Thus materialists help many people realise the gloom of the bigoted and traditional mentality and encourage them to seek the essence and real source of religion by referring to and adhering to the Qur'an. Although unintentionally, they obey God's command and serve His religion. Furthermore, they disclose all the simplicity of the mentality that presents a false religion invented in the name of God and proffered as Islam to all and they help weaken the sway of this bigoted system that threatens the bulk of society.

Thus willy-nilly and in accordance with their fate, they become the means whereby the decree of God about His upholding His true religion by causing the antagonists of religion counteract against each other is made true. God's law is stated in the Qur'an as follows;

> **And did not God check one set of people by means of another, the earth would indeed be full of mischief.** (Surat al-Baqara, 251)

At this point, we think it necessary to leave an open door for some advocates of the evolutionist materialist thought. These people might once have set out on an honest quest, yet have been driven away from the true religion under the influence of the idle talk produced in the name of Islam, falsehoods fabricated in the name of the Prophet, and hearsay stories to which they have been subject since their childhood and thus never have had the chance to discover the truth themselves. They might have learned religion from books by opponents of religion who try to identify Islam with falsehoods and fallacies that are not present in the Qur'an, and with the traditionalism and bigotry. The essence and origin of Islam are quite different and, moreover, completely incompatible with all that has been taught to them. For this reason, we suggest they get a Qur'an as soon as possible and read God's book with an open heart and a conscientious and unprejudiced view and learn the original religion from its true source. If they need assistance, they can refer to the books written by the author of this book, Harun Yahya, on the basic concepts in the Qur'an.

CHAPTER 15

Media: Fertile Ground for Evolution

As what we have examined so far has demonstrated, the theory of evolution rests on no scientific basis. However most people around the world are unaware of this and assume that evolution is a scientific fact. The biggest reason for this deception is the systematic indoctrination and propaganda conducted by the media about evolution. For this reason, we also have to mention the particular characteristics of this indoctrination and propaganda.

When we look at the Western media carefully, we frequently come across news dwelling on the theory of evolution. Leading media organisations, and well-known and "respectable" magazines periodically bring this subject up. When their approach is examined, one gets the impression that this theory is an absolutely proven fact leaving no room for discussion.

Ordinary people reading this kind of news naturally start to think that the theory of evolution is a fact as certain as any law of mathematics. News of this sort that appears in the prominent media engines is also picked up by local media. They print headlines in big fonts: "According to *Time* magazine, a new fossil that completes the gap in the fossil chain has been found"; or "*Nature*" indicates that scientists have shed light on the final issues of evolutionary theory". The finding of "the last missing link of the evolution chain" means nothing because there is not a single thing proven about evolution. Everything shown as evidence is false as we have described in the previous chapters. In addition to the media, the same holds true for scientific resources, encyclopaedias, and biology books.

In short, both the media and academic circles, which are at the disposal of anti-religionist power-centres, maintain an entirely evolutionist view and they impose this on society. This imposition is so effective that it has in time turned evolution into an idea that is never to be rejected. Denying evolution is seen as being contradictory to science and as disregarding fundamental realities. This is why, notwithstanding so many deficiencies

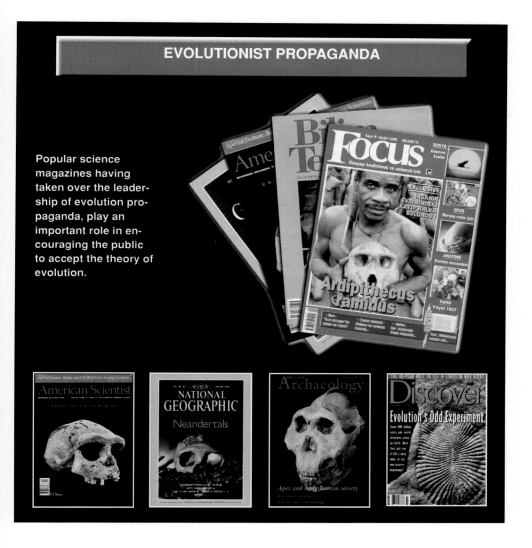

Popular science magazines having taken over the leadership of evolution propaganda, play an important role in encouraging the public to accept the theory of evolution.

that have so far been revealed (especially since the 1950s) and the fact that these have been confessed by evolutionist scientists themselves, today it is all but impossible to find any criticism of evolution in scientific circles or in the media.

Widely accepted as the most "respected" publishing vehicles on biology and nature in the West, magazines such as *Scientific American*, *Nature*, *Focus*, and *National Geographic* adopt the theory of evolution as an official ideology and try to present this theory as a proven fact.

FABLES FROM EVOLUTIONISTS

Evolution is, as once noted by a prominent scientist, a fairy tale for adults. It is a totally irrational and unscientific scenario, which suggests that non-living matter has some sort of a magical power and intelligence to create complex life forms. This long tale has some very interesting fables on some particular subjects. One of these curious evolutionary fables is the one about the "evolution of whale" that was published in *National Geographic*, widely respected as one of the most scientific and serious publications in the world:

> The Whale's ascendancy to sovereign size apparently began sixty million years ago when hairy, four-legged mammals, in search of food or sanctuary, ventured into water. As eons passed, changes slowly occurred. Hind legs disappeared, front legs changed into flippers, hair gave way to a thick smooth blanket of blubber, nostrils moved to the top of the head, the tail broadened into flukes, and in the buoyant water world the body became enormous.[1]

Besides the fact that there is not a single scientific basis for any of this, such an occurrence is also contrary to the principles of nature. This fable published in *National Geographic* is noteworthy for being indicative of the extent of the fallacies of seemingly serious evolutionist publications.

Another fable from evolutionists worth noting is on the origin of mammals. Evolutionists argue that mammals originated from a reptilian ancestor. But when it comes to explain the details of this alleged transformation, interesting narratives arise. Here is one of them:

> Some of the reptiles in the colder regions began to develop a method of keeping their bodies warm. Their heat output increased when it was cold and their heat loss was cut down when scales became smaller and more pointed, and evolved into fur. Sweating was also an adaptation to regulate the body temperature, a device to cool the body when necessary by evaporation of water. But incidentally the young of these reptiles began to lick the sweat of the mother for nourishment. Certain sweat glands began to secrete a richer and richer secretion, which eventually became milk. Thus the young of these early mammals had a better start in life.[2]

The idea that a well-designed food like milk could originate from sweat glands and all the other details above are just bizarre productions of evolutionary imagination, with no scientific basis.

1- Victor B. Scheffer, "Exploring the Lives of Whales", *National Geographic*, vol. 50, December 1976, p. 752
2- George Gamow, Martynas Ycas, *Mr. Tompkins Inside Himself*, London: Allen & Unwin, 1968, p. 149

Wrapped-up Lies

Evolutionists make great use of the advantage given to them by the "brain-washing" program of the media. Many people believe in evolution so unconditionally that they do not even bother to ask "how" and "why". This means that evolutionists can package their lies so as to be easily persuasive.

For instance, even in the most "scientific" evolutionist books the "transition from water to land", which is one of the greatest unaccounted-for phenomena of evolution, is "explained" with ridiculous simplicity. According to evolution, life started in water and the first developed animals were fish. The theory has it that one day these fish started to fling themselves on to the land for some reason or other, (most of the time, drought is said to be the reason), and the fish that chose to live on land, happened to have feet instead of fins, and lungs instead of gills.

Most evolutionist books do not tell the "how" of the subject. Even in the most "scientific" sources, the absurdity of this assertion is concealed behind sentences such as "the transfer from water to land was achieved".

How was this "transfer" achieved? We know that a fish cannot live for more than a few minutes out of water. If we suppose that the alleged drought occurred and the fish had to move towards the land, what would have happened to the fish? The response is evident. All of the fish coming out of the water would die one by one in a few minutes. Even if this process had had lasted for a period of ten million years, the answer would still be the same: fish would die one by one. The reason is that such a complex organ as a complete lung cannot come into being by a sudden "accident", that is, by mutation; but half a lung, on the other hand, is of no use at all.

But this is exactly what the evolutionists propose. "**Transfer from water to land**", "**transfer from land to air**" and many more alleged leaps are "explained" in these illogical terms. As for the formation of really complex organs such as the eye and ear, evolutionists prefer not to say anything at all.

It is easy to influence the man on the street with the package of "science". You draw an imaginary picture representing transfer from water to land, you invent Latin words for the animal in the water, its "descendant" on land, and the "transitional intermediary form" (which is an imaginary animal), and then fabricate an elaborate lie: "*Eusthenopteron* transformed first into *Rhipitistian Crossoptergian*, then *Ichthyostega* in a long evolutionary process". If you put these words in the mouth of a scientist with thick glasses and a white coat, you would succeed in convincing many people, because the media, which dedicates itself to promoting evolution, would announce the good news to the world with great enthusiasm.

CHAPTER 16

Conclusion: Evolution Is a Deceit

There is much other evidence, as well as scientific laws, invalidating evolution, but in this book we have only been able to discuss some of them. Even those should be enough to reveal a most important truth: Although it is cloaked in the guise of science, the theory of evolution is nothing but a deceit: a deceit defended only for the benefit of materialistic philosophy; a deceit based not on science but on brainwashing, propaganda, and fraud.

We can summarise what we have noted so far as follows:

The Theory of Evolution has Collapsed

The theory of evolution is a theory that fails at the very first step. The reason is that evolutionists are unable to explain even the formation of a single protein. Neither the laws of probability nor the laws of physics and chemistry offer any chance for the fortuitous formation of life.

Does it sound logical or reasonable when not even a single chance-formed protein can exist, that millions of such proteins combined in an order to produce the cell of a living thing; and that billions of cells managed to form and then came together by chance to produce living things; and that from them generated fish; and that those that passed to land turned into reptiles, birds, and that this is how all the millions of different species on earth were formed?

Even if it does not seem logical to you, evolutionists do believe this fable.

However, it is merely a belief-or rather a faith-because they do not have even a single piece of evidence to verify their story. They have never found a single transitional form such as a half-fish/half-reptile or half-reptile/half-bird. Nor have they been able to prove that a protein, or even a single amino acid molecule composing a protein, could have formed under what they call primordial earth conditions; not even in their elaborately-equipped laboratories have they succeeded in doing that. On the contrary, with their every effort, evolutionists themselves have demonstrated that no evolutionary process has ever occurred nor could ever have occurred at any time on earth.

Evolution Can Not Be Verified in the Future Either

Seeing this, evolutionists can only console themselves by dreaming that science will somehow resolve all these dilemmas in time. However, that science should ever verify such an entirely groundless and illogical claim is out of the question no matter how many years may pass by. On the contrary, as science progresses it only makes the nonsense of evolutionists' claims clearer and plainer.

That is how it has been so far. As more details on the structure and functions of the living cell were discovered, it became abundantly clear that the cell is not a simple, randomly-formed composition, as was thought to be the case according to the primitive biological understanding of Darwin's time.

With the situation being so self-evident, denying the fact of creation and basing the origins of life on extremely unlikely coincidences, and then defending these claims with insistence, may later become a source of great humiliation. As the real face of the evolution theory comes more and more into view and as public opinion comes to see the truth, it may not be long before the purblind fanatic advocates of evolution will not be able to show their faces.

The Biggest Obstacle to Evolution: Soul

There are many species in the world that resemble one another. For instance, there may be many living beings resembling a horse or a cat and many insects may look like one another. These similarities do not surprise anyone.

The superficial similarities between man and ape somehow attract too much attention. This interest sometimes goes so far as to make some people believe the false thesis of evolution. As a matter of fact, the superficial similarities between men and apes do signify nothing. The rhinoceros beetle and the rhinoceros also share certain superficial resemblances but it would be ludicrous to seek to establish some kind of an evolutionary link between these two creatures, one being an insect and the other a mammal, on the grounds of that resemblance.

Other than superficial similarity, apes cannot be said to be closer to man than to other animals. Actually, if level of intelligence is considered, then the honeybee producing the geometrically miraculous structure of the honeycomb or the spider building up the engineering miracle of the spider web can be said to be closer to man. They are even superior in some aspects.

There is a very big difference between man and ape regardless of a mere outward resemblance. An ape is an animal and is no different from a horse or a dog considering its level of consciousness. Yet man is a conscious, strong-willed being that can think, talk, understand, decide, and judge. All of these features are the functions of the soul that man possesses. The soul is the most important difference that interposes a huge gap between man and other creatures. No physical similarity can close this gap between man and any other living being. In nature, the only living thing that has a soul is man.

God Creates According to His Will

Would it matter if the scenario proposed by evolutionists really had taken place? Not a bit. The reason is that each stage advanced by evolutionary theory and based on coincidence could only have occurred as a result of a miracle. Even if life did come about gradually through such a succession of stages, each progressive stage could only have been brought about by a conscious will. It is not just implausible that those stages could have occurred by chance, it is impossible.

If is said that a protein molecule had been formed under the primordial atmospheric conditions, it has to be remembered that it has been already demonstrated by the laws of probability, biology, and chemistry that this could not have been by chance. But if it must be posited that it was produced, then there is no alternative but to admit that it owed its existence to the will of a Creator. The same logic applies to the entire hypothesis put forward by evolutionists. For instance, there is neither paleontological evidence nor a physical, chemical, biological, or logical justification proving that fish passed from water to land and formed the land animals. But if one must have it that fish clambered onto the land and turned into reptiles, the maker of that claim should also accept the exis-

Conclusion: Evolution Is a Deceit

tence of a Creator capable of making whatever He wills come into being with the mere word "be". Any other explanation for such a miracle is inherently self-contradictory and a violation of the principles of reason.

The reality is clear and evident. All life is the product of a perfect design and a superior creation. This in turn provides concrete evidence for the existence of a Creator, the Possessor of infinite power, knowledge, and intelligence.

That Creator is God, the Lord of the heavens and of the earth, and of all that is between them.

CHAPTER 17

The Fact of Creation

In the previous sections of the book, we examined why the Theory of Evolution, which proposes that life was not created, is a fallacy completely contrary to scientific facts. We saw that modern science has revealed a very explicit fact through certain branches of science such as paleontology, biochemistry, and anatomy. This fact is that God creates all living beings.

In fact, to notice this fact one does not necessarily need to appeal to the complicated results obtained in biochemistry laboratories or geological excavations. The signs of an extraordinary wisdom are discernible in whatever living being one observes. There is a great technology and design in the body of an insect or a tiny fish in the depths of the sea never attained by human beings. Some living beings which even do not have a brain perfectly perform so complicated tasks as not to be accomplished even by human beings.

This great wisdom, design and plan that prevails overall in nature, provides solid evidence for the existence of a supreme Creator dominating over the whole of nature, and this Creator is God. God has furnished all living beings with extraordinary features and showed men the evident signs of His existence and might.

In the following pages, we will examine only a few of the countless evidences of Creation in nature.

Honey Bees and the Architectural Wonders of Honeycombs

Bees produce more honey than they actually need and store it in honeycombs. The hexagonal structure of the honeycomb is well-known to everyone. Have you ever wondered why bees construct hexagonal honeycombs rather than octagonal, or pentagonal?

Mathematicians looking for answer to this question reached an interesting conclusion: "A hexagon is the most appropriate geometric form for the maximum use of a given area."

The Fact of Creation

A hexagonal cell requires the minimum amount of wax for construction while it stores the maximum amount of honey. So the bee uses the most appropriate form possible.

The method used in the construction of the honeycomb is also very amazing: bees start the construction of the hive from two-three different places and weave the honeycomb simultaneously in two-three strings. Though they start from different places, the bees, great in number, construct identical hexagons and then weave the honeycomb by combining these together and meeting in the middle. The junction points of the hexagons are assembled so deftly that there is no sign of their being subsequently combined.

In the face of this extraordinary performance, we, for sure, have to admit the existence of a superior will that ordains these creatures. Evolutionists want to explain away this achievement with the concept of "instinct" and try to present it as a simple attribute of the bee. However, if there is an instinct at work, if this rules over all bees and provides that all bees work in harmony though uninformed of one another, then it means that there is an exalted Wisdom that rules over all these tiny creatures.

To put it more explicitly, God, the creator of these tiny creatures, "inspires" them with what they have to do. This fact was declared in the Qur'an fourteen centuries ago:

And your Sustainer has inspired the honey bee: "Prepare for yourself dwellings in mountains and in trees, and in what (men) build; and then eat of all manner of fruit, and find with skill the spacious paths of your Sus-

tainer". There issues from within their bodies a drink of varying colours, wherein is healing for men: verily in this is a Sign for those who give thought. (Surat an-Nahl, 68-69)

Amazing Architects: Termites

No one can help being taken by surprise upon seeing a termite nest erected on the ground by termites. This is because the termite nests are architectural wonders that rise up as high as 5-6 meters. Within this nest are sophisticated systems to meet all the needs of termites that can never appear in sunlight because of their body structure. In the nest, there are ventilation systems, canals, larva rooms, corridors, special fungus production yards, safety exits, rooms for hot and cold weather; in brief, everything. What is more astonishing is that the termites which construct these wondrous nests are blind.[185]

Despite this fact, we see, when we compare the size of a termite and its nest, that termites successfully overcome an architectural project by far 300 times bigger than themselves.

Termites have yet another amazing characteristic: if we divide a termite nest into two in the first stages of its construction, and then reunite it after a certain while, we will see that all passage-ways, canals and roads intersect with each other. Termites carry on with their task as if they were never separated from each other and ordained from a single place.

The Woodpecker

Everyone knows that woodpeckers build their nests by pecking tree trunks. The point many people do not consider is how woodpeckers undergo no brain haemorrhage when they so strongly tattoo with their head. What the woodpecker does is in a way similar to a human driving a nail in the wall with his head. If a human ventured to do something like that, he would probably undergo a brain shock followed by a brain haemorrhage. A woodpecker, however, can peck a hard tree trunk 38-43 times between 2.10 and 2.69 seconds and nothing happens to it.

Nothing happens because the head structure of woodpeckers are created as fit for this job. The woodpecker's skull has a "suspension" system that reduces and absorbs the force of the strokes. There are special softening tissues between the bones in its skull.[186]

The Sonar System of Bats

Bats fly in pitch dark without trouble and they have a very interesting navigation system to do this. It is what we call "sonar" system, a system whereby the shapes of the surrounding objects are determined according to the echo of the sound waves.

A young person can barely detect a sound with a frequency of 20,000 vibrations per second. A bat furnished with a specially designed "sonar system", however, makes use of sounds having a frequency of between 50,000 and 200,000 vibrations per second. It sends these sounds in all directions 20 or 30 times each second. The echo of the sound is so powerful that the bat not only understands the existence of objects in its path, but also detects the location of its swift-flying prey.[187]

Whales

Mammals regularly need to breathe and for this reason water is not a very convenient environment for them. In a whale, which is a sea mammal, however, this problem is handled with a breathing system far more efficient than that of many land-dwelling animals. Whales breathe out one at a time discharging 90% of the air they use. Thus, they need to breathe only at very long intervals. At the same time, they have a highly concentrated substance called "myoglobin" that helps them store oxygen in their muscles. With the help of these systems, finback whale, for instance, can dive as deep as 500 meters and swim for 40 minutes without breathing at all.[188] The nostrils of the whale, on the other hand, are placed on its back unlike land-dwelling mammals so that it can easily breathe.

The Design in The Gnat

We always think of the gnat as a flying animal. In fact, the gnat spends its developmental stages under water and gets out from under water through an exceptional "design" being provided with all the organs it needs.

The gnat starts to fly with special sensing systems at its disposal to detect the place of its prey. With these systems, it resembles a war plane loaded with detectors of heat, gas, dampness and odour. It even has an ability to "see in conformity with the temperature" that helps it find its prey even in pitch dark.

The "blood-sucking" technique of the gnat comes with an incredibly complex system. With its six-bladed cutting system, it cuts the skin like a saw. While the cutting process goes on, a secretion secreted on the wound benumbs the tissues and the person does not even realise that his blood is being sucked. This secretion, at the same time, prevents the clotting of the blood and secures the continuance of the sucking process.

With even one of these elements missing, the gnat will not be able to feed on blood and carry on its generation. With its exceptional design, even this tiny creature is an evident sign of Creation on its own.

In the Qur'an, the gnat is accentuated as an example displaying the existence of God to the men of understanding:

> **Surely God disdains not to set forth any parable - (that of) a (female) gnat or any thing above that; then as for those who believe, they know that it is the truth from their Lord, and as for those who disbelieve, they say: What is it that God means by this parable: He causes many to err by it and many He leads aright by it! but He does not cause to err by it (any) except the transgressors, (Surat al-Baqara, 26)**

Hunting Birds with Keen Eyesight

Hunting birds have keen eyes that enable them to make perfect distance adjustments while they attack their prey. In addition their large eyes contain more vision cells, which means better sight. There are more than one million vision cells in the eye of a hunting bird.

Eagles that fly at thousands of meters high have such sharp eyes that they can scan the earth perfectly at that distance. Just as war planes detect their targets from thousands of meters away, so do eagles spot their prey, perceiving the slightest colour shift or the slightest movement on the earth. The eagle's eye has an angle of vision of three hundred degrees and it can magnify a given image around six to eight times. Eagles can scan an area of 30,000 hectares while flying 4,500 meters above it. They can easily distinguish a rabbit hidden among grasses from an altitude of 1,500 meters. It is evident that this extraordinary eye structure of the eagle is specially designed for this creature.

The Thread of the Spider

The spider named Dinopis has a great skill for hunting. Rather than weaving a static web and waiting for its prey, it weaves a small yet highly unusual web that it throws on its prey. Afterwards, it tightly wraps up its prey with this web. The entrapped insect can do nothing to extricate itself. The web is so perfectly constructed that the insect gets even more entangled as it gets more alarmed. In order to store its food, the spider wraps the prey with extra strands, almost as if it were packaging it.

How does this spider make a web so excellent in its mechanical design and chemical structure? It is impossible for the spider to have acquired such a skill by coincidence as is claimed by evolutionists. The spider is devoid of faculties such as learning and memorising and does not have even a brain to perform these things. Obviously, this skill is bestowed on the spider by its creator, God, Who is Exalted in Power.

Very important miracles are hidden in the thread of the spiders. This thread, with a diameter of less than one thousandth of a millimetre, is 5 times stronger than a steel wire having the same thickness. This thread has yet another characteristic of being extremely light. A length of this thread long enough to encircle the world would weigh only 320 grams.[189] Steel, a substance specially produced in industrial works, is one of the strongest materials manufactured by mankind. However, the spider can produce in its body a far firmer thread than steel. While man produces steel, he makes use of his centuries-old knowledge and technology; which knowledge or technology, then, does the spider use while producing its thread?

As we see, all technological and technical means at the disposal mankind lag behind those of a spider.

Hibernating Animals

Hibernating animals can go on living although their body temperature falls to the same degree as the cold temperature outside. How do they manage this?

Mammals are warm-blooded. This means that under normal conditions, their body temperature always remains constant because the natural thermostat in their body keeps on regulating this temperature. However, during hibernation, the normal body heat of small mammals, like the squirrel rat with a normal body heat of 40 degrees, drops down to a little bit above the freezing point as if adjusted by some kind of a key. The body metabolism slows down to a great extent. The animal starts breathing very slowly and its normal heartbeat, which is 300 times a minute, falls to 7-10 beats a minute.

Its normal body reflexes stop and the electrical activities in its brain slow down almost to undetectability.

One of the dangers of motionlessness is the freezing of tissues in very cold weather and their being destroyed by ice crystals. Hibernating animals however are protected against this danger thanks to the special features they are endowed with. The body fluids of hibernating animals are retained by chemical materials having high molecular masses. Thus, their freezing point is decreased and they are protected from harm.[190]

Electrical Fish

Certain species of some fish types such as electric eel and electric ray utilise the electricity produced in their bodies either to protect themselves from their enemies or to paralyse their prey. In every living being - including man - is a little amount of electricity. Man, however, cannot direct this electricity or take it under control to use it for his own benefit. The above-mentioned creatures, on the other hand, have an electrical current as high as 500-600 volts in their bodies and they are able use this against their enemies. Furthermore, they are not adversely affected by this electricity.

The energy they consume to defend themselves is recovered after a certain time like the charging of a battery and electrical power is once again ready for use. Fish do not use the high-voltage electricity in their small bodies only for defence purposes. Besides providing the means for finding their way in deep dark waters, electricity also helps them sense objects without seeing them. Fish can send signals by using the electricity in their bodies. These electric signals reflect back after hitting solid objects and these reflections give the fish information about the object. This way, fish can determine the distance and size of the object.[191]

An Intelligent Plan on Animals: Camouflage

One of the features that animals possess in order to keep living is the art of hiding themselves-that is, "camouflage".

Animals feel the necessity of hiding themselves for two main reasons: for hunting and for protecting themselves from predators. Camouflage differs from all other methods with its particular involvement of utmost intelligence, skill, aesthetics and harmony.

The camouflage techniques of animals are truly amazing. It is almost

Above: Tree louse imitating tree thorns. Right above: A snake concealing itself by suspending itself among leaves. Right below: A caterpillar settled right in the middle of a leaf to go unnoticed.

impossible to identify an insect that is hidden in a tree trunk or another creature hidden under a leaf.

Leaf louse that suck the juices of plants feed themselves on plant stalks by pretending to be thorns. By this method, they aim to trick birds, their biggest enemies, and ensure that birds will not perch on these plants.

Cuttlefish

Under the skin of the cuttlefish is arrayed a dense layer of elastic pigment sacs called chromatophores. They come mainly in yellow, red, black and brown. At a signal, the cells expand and flood the skin with the appropriate shade. That is how the cuttlefish takes on the colour of the rock it stands on and makes a perfect camouflage.

This system operates so effectively that the cuttlefish can also create a complex zebra-like striping.[192]

Left: A cuttlefish that makes itself look like the sandy surface. **Right:** The bright yellow colour the same fish turns in case of danger, such as when it is seen by a diver.

Different Vision Systems

For many sea-dwelling animals, seeing is extremely important for hunting and defence. Accordingly, most of the sea-dwelling animals are equipped with eyes perfectly designed for underwater.

Under water, the ability to see becomes more and more limited with depth, especially after 30 meters. Organisms living at this depth, however, have eyes created according to the given conditions.

Sea-dwelling animals, unlike land-dwelling animals, have spherical lenses in perfect accordance with the needs of the density of the water they inhabit. Compared to the wide elliptical eyes of land-dwelling animals, this spherical structure is more serviceable for sight under water; it is adjusted to see objects in close-up. When an object at a greater distance is focused upon, the whole lens system is pulled backwards by the help of a special muscle mechanism within the eye.

One other reason why the eyes of the fish are spherical is the refraction of light in water. Because the eye is filled with a liquid having almost the same density as water, no refraction occurs while an image formed outside is reflected on the eye. Consequently, the eye lens fully focuses the image of the outside object on the retina. The fish, unlike human beings, sees very sharply in water.

Some animals like octopus have rather big eyes to compensate for the poor light in the depths of water. Below 300 meters, big-eyed fish need to capture the flashes of the surrounding organisms to notice them. They have to be especially sensitive to the feeble blue light penetrating into the water. For this reason, there are plenty of sensitive blue cells in the retina of their eyes.

As is understood from these examples, every living being has distinctive eyes specially designed to meet its particular needs. This fact proves that they are all created just the way they have to be by a Creator Who has eternal wisdom, knowledge and power.

Special Freezing System

A frozen frog embodies an unusual biological structure. It shows no signs of life. Its heartbeat, breathing and blood circulation have come completely to a halt. When the ice melts, however, the same frog returns to life as if it is has woken up from sleep.

Normally, a living being in the state of freezing confronts many fatal risks. The frog, however, does not face any of them. It has the main feature of producing plenty of glucose while it is in that state. Just like a diabetic, the blood sugar level of the frog reaches very high levels. It can sometimes go as high as 550 milimol/liter. (This figure is normally between 1-5 mmol/litre for frogs and 4-5 mmol/litre for human body). This extreme glucose concentration may cause serious problems in normal times.

In a frozen frog, however, this extreme glucose keeps water from leaving cells and prevents shrinkage. The cell membrane of the frog is highly permeable to glucose so that glucose finds easy access to cells. The high level of glucose in the body reduces the freezing temperature causing only a very small amount of the animal's inner body liquid to turn to ice in the cold. Research has showed that glucose can feed frozen cells as well. During this period, besides being the natural fuel of the body, glucose also stops many metabolic reactions like urea synthesis and thus prevents different food sources of the cell from being exhausted.

How does such a high amount of glucose in the frog's body come about all of a sudden? The answer is quite interesting: this living being is equipped with a very special system in charge of this task. As soon as ice appears on the skin, a message travels to the liver making the liver convert some of its stored glycogen into glucose. The nature of this message travelling to the liver is still unknown. Five minutes after the message is received, the sugar level in the blood steadily starts to increase.[193]

Unquestionably the animal's being equipped with a system that entirely changes its metabolism to meet all of its needs just when it is required can only be possible through the flawless plan of the All-Mighty Creator. No coincidence can generate such a perfect and complex system.

Albatrosses

Migratory birds minimise energy consumption by using different "flight techniques". Albatrosses are also observed to have such a flight style. These birds, which spend 92% of their lives on the sea, have wing spans of up to 3,5 meters. The most important characteristic of albatrosses is their flight style: they can fly for hours without beating their wings at all. To do so, they glide along in the air keeping their wings constant by making use of the wind.

It requires a great deal of energy to keep wings with a wing span of 3.5 meters constantly open. Albatrosses, however, can stay in this position for hours. This is due to the special anatomical system they are bestowed with from the moment of their birth. During flight, the wings of the albatross are blocked. Therefore, it does not need to use any muscular power. Wings are lifted only by muscle layers. This greatly helps the bird during its flight. This system reduces the energy consumed by the bird during flight. The albatross does not use energy because it does not beat its wings or waste energy to keep its wings outstretched. Flying for hours by making exclusive use of wind provides an unlimited energy source for it. For instance, a 10-kilo-albatross loses only 1% of its body weight while it travels for 1,000 kms. This is indeed a very small rate. Men have manufactured gliders taking albatrosses as a model and by making use of their fascinating flight technique.[194]

An Arduous Migration

Pacific salmon have the exceptional characteristic of returning to the rivers in which they hatched to reproduce. Having spent part of their lives in the sea, these animals come back to fresh water to reproduce.

When they start their journey in early summer, the colour of the fish is bright red. At the end of their journey, however, their colour turns black. At the outset of their migration, they first draw near to the shore and try to reach rivers. They perseveringly strive to go back to their birthplace. They reach the place where they hatched by leaping over turbulent rivers, swimming upstream, surmounting waterfalls and dykes. At the end of this 3,500-4,000 km. journey, female salmon readily have eggs just as male salmons have sperm. Having reached the place where they hatched, female salmon lay around 3 to 5 thousand eggs as male salmon fertilise them. The fish suffer much damage as a result of this migration and hatching period. Females that lay eggs become exhausted; their tail fins are worn down and their skin starts to turn black. The same is true also for males. The river soon overflows with dead salmon. Yet another salmon generation is ready to hatch out and make the same journey.

How salmon complete such a journey, how they reach the sea after they hatch, and how they find their way are just some of the questions that remain to be answered. Although many suggestions are made, no definite solution has yet been reached. What is the power that makes salmon undertake a return of thousands of kilometres back to a place unknown to them? It is obvious that there is a superior Will ruling over and controlling all these living beings. It is God, the Sustainer of all the worlds.

Koalas

The oil found in eucalyptus leaves is poisonous to many mammals. This poison is a chemical defence mechanism used by eucalyptus trees against their enemies. Yet there is a very special living being that gets the better of this mechanism and feeds on poisonous eucalyptus leaves: a marsupial called the koala. Koalas make their homes in eucalyptus trees while they also feed on them and obtain their water from them.

Like other mammals, koalas also cannot digest the cellulose present in the trees. For this, it is dependent on cellulose-digesting micro-organisms. These micro-organisms are heavily populated in the convergence point of small and large intestines, the caecum which is the rear extension of the intestinal system. The caecum is the most interesting part of the digestion system of the koala. This segment functions as a fermentation chamber where microbes are made to digest cellulose while the passage of the leaves is delayed. Thus, the koala can neutralise the poisonous effect of the oils in the eucalyptus leaves.[195]

Hunting Ability in Constant Position

The South African sundew plant entraps insects with its viscous hairs. The leaves of this plant are full of long, red hairs. The tips of these hairs are covered with a fluid that has a smell that attracts insects. Another feature of the fluid is its being extremely viscous. An insect that makes its way to the source of the smell gets stuck in these viscous hairs. Shortly afterwards the whole leaf is closed down on the insect that is already entangled in the hairs and the plant extracts the protein essential for itself from the insect by digesting it.[196]

The endowment of a plant with no possibility of moving from its place with such a faculty is no doubt the evident sign of a special design. It is impossible for a plant to have developed such a hunting style out of its own consciousness or will, or by way of coincidence. So, it is all the more impossible to overlook the existence and might of the Creator Who has furnished it with this ability.

The Design In Bird Feathers

At first glance, bird feathers seem to have a very simple structure. When we study them closer, however, we come across the very complex structure of feathers that are light yet extremely strong and waterproof.

Birds should be as light as possible in order to fly easily. The feathers are

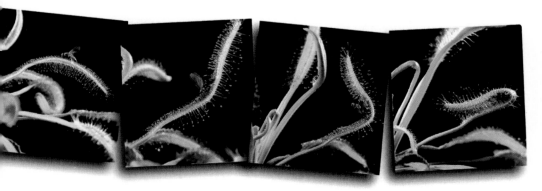

Left: An open Sundew. Right: A closed one.

made up of keratin proteins keeping with this need. On both sides of the stem of a feather are veins and on each vein are around 400 tiny barbs. On these 400 barbs are a total of tinier 800 barbs, two on each. Of the 800 tinier barbs which are crowded on a small bird feather, those located towards the front part have another 20 barbs on each of them. These barbs fasten two feathers to one another just like two pieces of cloth tacked up on each other. In a single feather are approximately 300 million tiny barbs. The total number of barbs in all the feathers of a bird is around 700 billion.

There is a very significant reason for the bird feather being firmly interlocked with each other with barbs and clasps. The feathers should hold tightly on the bird so as not to fall out in any movement whatsoever. With the mechanism made up of barbs and clasps, the feathers hold so tightly on the bird that neither strong wind, nor rain, nor snow cause them to fall out.

Furthermore, the feathers in the abdomen of the bird are not the same as the feathers in its wings and tail. The tail is made up of relatively big feathers to function as rudder and brakes; wing feathers are designed so as to expand the area surface during the bird's wing beating and thus increase the lifting force.

Basilisk: The Expert of Walking on Water

Few animals are able to walk on the surface of water. One such rarity is basilisk, which lives in Central America and is seen below. On the sides of the toes of basilisk's hind feet are flaps that enable them to splash water.

The basilisk lizard is one of those rare animals that can move establishing a balance between water and air.

These are rolled up when the animal walks on land. If the animal faces danger, it starts to run very fast on the surface of a river or a lake. Then the flaps on its hind feet are opened and thus more surface area is provided for it to run on water.[197]

This unique design of basilisk is one of the evident signs of conscious Creation.

Photosynthesis

Plants unquestionably play a major role in making the universe a habitable place. They clean the air for us, keep the temperature of the planet at a constant level, and balance the proportions of gases in the atmosphere. The oxygen in the air we breathe is produced by plants. An important part of our food is also provided by plants. The nutritional value of plants comes from the special design in their cells to which they also owe their other features.

The plant cell, unlike human and animal cells, can make direct use of solar energy. It converts the solar energy into chemical energy and stores it in nutrients in very special ways. This process is called "photosynthesis". In fact, this process is carried out not by the cell but by chloroplasts, organelles that give plants their green colour. These tiny green organelles only observable by microscope are the only laboratories on earth that are capable of storing solar energy in organic matter.

The amount of matter produced by plants on the earth is around 200 billion tons a year. This production is vital to all living things on the earth.

The production made by plants is realised through a very complicated chemical process. Thousands of "chlorophyll" pigments found in the chloroplast react to light in an incredibly short time, something like one thousandth of a second. This is why many activities taking place in the chlorophyll have still not been observed.

Converting solar energy into electrical or chemical energy is a very recent technological breakthrough. In order to do this, high-tech instruments are used. A plant cell so small as to be invisible to the naked human eye has been performing this task for millions of years.

This perfect system displays Creation once more for all to see. The very complex system of photosynthesis is a consciously-designed mechanism that God creates. A matchless factory is squeezed in a minuscule unit area in the leaves. This flawless design is only one of the signs revealing that God, the Sustainer of all worlds, creates all living things.

PART II

THE REFUTATION OF MATERIALISM

WARNING !

The chapter you are now about to read reveals a crucial secret of your life. You should read it very attentively and thoroughly for it is concerned with a subject that is liable to make a fundamental change in your outlook to the external world. The subject of this chapter is not just a point of view, a different approach, or a traditional philosophical thought: it is a fact which everyone, believing or unbelieving, must admit and which is also proven by science today.

CHAPTER 18

The Real Essence of Matter

People who contemplate their surroundings conscientiously and wisely realise that everything in the universe-both living and non-living-must have been created. So the question becomes that of "Who is the creator of all these things?"

It is evident that "**the fact of creation**", which reveals itself in every aspect of the universe, cannot be an outcome of the universe itself. For example, a bug could not have created itself. The solar system could not have created or organised itself. Neither plants, humans, bacteria, erythrocytes (red-blood corpuscles), nor butterflies could have created themselves. The possibility that these all could have originated "by chance" is not even imaginable.

We therefore arrive at the following conclusion: Everything that we see has been created. But nothing that we see can be "creators" themselves. The Creator is different from and superior to all that we see with our eyes, a superior power that is invisible but whose existence and attributes are revealed in everything that exists.

This is the point at which those who deny the existence of God demur. These people are conditioned not to believe in His existence unless they see Him with their eyes. These people, who disregard the fact of "creation", are forced to ignore the actuality of "**creation**" manifested all throughout the universe and falsely prove that the universe and the living things in it have not been created. Evolutionary theory is a key example of their vain endeavours to this end.

The basic mistake of those who deny God is shared by many people who in fact do not really deny the existence of God but have a wrong perception of Him. They do not deny creation, but have superstitious beliefs about "where" God is. Most of them think that God is up in the "sky". They tacitly imagine that God is behind a very distant planet and interferes with "worldly affairs" once in a while. Or perhaps that He does not intervene at all: He created the universe and then left it to itself and people are left to determine their fates for themselves.

Still others have heard that in the Qur'an it is written that God is "everywhere" but they cannot perceive what this exactly means. They tacitly think that God surrounds everything like radio waves or like an invisible, intangible gas.

However, this notion and other beliefs that are unable to make clear **"where" God is** (and maybe deny Him because of that) are all based on a common mistake. They hold a prejudice without any grounds and then are moved to wrong opinions of God. What is this prejudice?

This prejudice is about the nature and characteristics of matter. We are so conditioned to suppositions about the existence of matter that we never think about whether or not it does exist or is only created as an image. Modern science demolishes this prejudice and discloses a very important and imposing reality. In the following pages, we will try to explain this great reality to which the Qur'an points.

The World Of Electrical Signals

All the information that we have about the world we live in is conveyed to us by our five senses. The world we know of consists of what our eye sees, our hand feels, our nose smells, our tongue tastes, and our ear hears. We never think that the "external" world can be other than what our senses present to us as we have been dependent only on those senses since the day of our birth.

Modern research in many different fields of science however points to a very different understanding and creates serious doubt about our senses and the world that we perceive with them.

The starting-point of this approach is that the notion of an "external world" shaped in our brain is only a response formed in our brain by electrical signals. The redness of the apple, the hardness of the wood, moreover, your mother, father, your family, and everything that you own, your house, your job, and the lines of this book, are comprised only of electrical signals.

Frederick Vester explains the point that science has reached on this subject:

The Real Essence of Matter

Stimulations coming from an object are converted into electrical signals and cause an effect in the brain. When we "see", we in fact view the effects of these electrical signals in our mind.

Statements of some scientists posing that "man is an image, everything experienced is temporary and deceptive, and this universe is a shadow", seems to be proven by science in our day.[198]

The famous philosopher George Berkeley's comment on the subject is as follows:

> We believe in the existence of objects just because we see and touch them, and they are reflected to us by our perceptions. However, our perceptions are only ideas in our mind. Thus, objects we captivate by perceptions are nothing but ideas, and these ideas are essentially in nowhere but our mind... Since all these exist only in the mind, then it means that we are beguiled by deceptions when we imagine the universe and things to have an existence outside the mind. So, none of the surrounding things have an existence out of our mind.[199]

In order to clarify the subject, let us consider our sense of sight, which provides us with the most extensive information about the external world.

How Do We See, Hear, And Taste?

The act of seeing is realised in a very progressive way. Light clusters (photons) that travel from the object to the eye pass through the lens in front of the eye where it is broken and falls reversely on the retina at the back of the eye. Here, the impinging light is turned into electrical signals that are transmitted by neurons to a tiny spot called the centre of vision in the back part of the brain. This electrical signal is perceived as an image in this centre in the brain after a series of processes. The act of seeing actually takes place in this tiny spot at the posterior part of the brain which is **pitch-dark and completely insulated from light**.

Now, let us reconsider this seemingly ordinary and unremarkable process. When we say that "we see", we are in fact seeing the effects of the impulses reaching our eye and induced in our brain after they are transformed into electrical signals. That is, **when we say that "we see", we are actually observing electrical signals in our mind.**

All the images we view in our lives are formed in our centre of vision, which makes up only a few cubic centimetres of the volume of the brain. Both the book you are now reading and the boundless landscape you see when you gaze at the horizon fit into this tiny space. Another point that has to be kept in mind is that as we have noted before, the brain is insulated from light; its inside is absolutely dark. The brain has no contact with light itself.

We can explain this interesting situation with an example. Let us suppose that there is a burning candle in front of us. We can sit across from this candle and watch it at length. However, during this period of time, our brain never has any direct contact with the candle's original light. Even as we see the light of the candle, the inside of our brain is solid dark. We watch a colourful and bright world inside our dark brain.

R.L. Gregory makes the following explanation about the miraculous aspect of seeing, an action that we take so very much for granted:

> We are so familiar with seeing, that it takes a leap of imagination to realise that there are problems to be solved. But consider it. We are given tiny distorted upside-down images in the eyes, and we see separate solid objects in surrounding space. From the patterns of simulation on the retinas we perceive the world of objects, and **this is nothing short of a miracle.**[200]

The same situation applies to all our other senses. Sound, touch, taste and smell are all transmitted to the brain as electrical signals and are perceived in the relevant centers in the brain.

The sense of hearing takes place in the same manner. The outer ear picks up available sounds by the auricle and directs them to the middle ear; the middle ear transmits the sound vibrations to the inner ear by intensifying them; the inner ear sends these vibrations to the brain by translating them into electrical signals. Just as with the eye, the act of hearing finalises in the centre of hearing in the brain. The brain is insulated from

The Real Essence of Matter

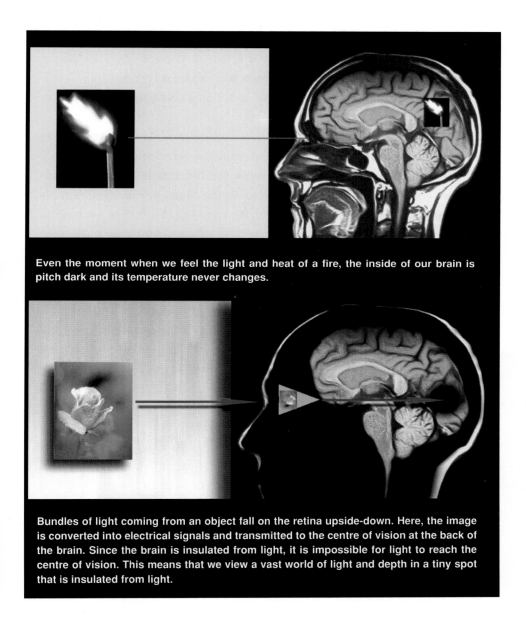

Even the moment when we feel the light and heat of a fire, the inside of our brain is pitch dark and its temperature never changes.

Bundles of light coming from an object fall on the retina upside-down. Here, the image is converted into electrical signals and transmitted to the centre of vision at the back of the brain. Since the brain is insulated from light, it is impossible for light to reach the centre of vision. This means that we view a vast world of light and depth in a tiny spot that is insulated from light.

sound just like it is from light. Therefore, no matter how noisy it is outside, the inside of the brain is completely silent.

Nevertheless, even the subtlest sounds are perceived in the brain. This is such a precision that the ear of a healthy person hears everything without any atmospheric noise or interference. In your brain, which is insulated from sound, you listen to the symphonies of an orchestra, hear all the noises in a crowded place, and perceive all the sounds within a wide frequency ranging from the rustling of a leaf to the roar of a jet plane. However, if the sound level in your brain were to be measured by a sensitive device at that moment, it would be seen that a complete silence is prevailing there.

Our perception of odour forms in a similar way. Volatile molecules emitted by things such vanilla or a rose reach the receptors in the delicate hairs in the epithelium region of the nose and become involved in an interaction. This interaction is transmitted to the brain as electrical signals and perceived as smell. Everything that we smell, be it nice or bad, is nothing but the brain's perceiving of the interactions of volatile molecules after they have been transformed into electrical signals. You perceive the scent of a perfume, a flower, a food that you like, the sea, or other odors you like or dislike in your brain. The molecules themselves never reach the brain. Just as with sound and vision, what reaches your brain is simply electrical signals. In other words, all the odours that you have assumed to belong to external objects since you were born are just electrical signals that you feel through your sense organs.

We see everything around us as coloured inside the darkness of our brains, just as this garden looks coloured from the window of a darkened room.

Similarly, there are four different types of chemical receptors in the front part of a human's tongue. These pertain to the tastes of salty, sweet, sour, and bitter. Our taste receptors transform these perceptions into electrical signals after a chain of chemical processes and transmit them to the brain. These signals are perceived as taste by the brain. The taste you get when you eat a chocolate bar or a fruit that

The Real Essence of Matter

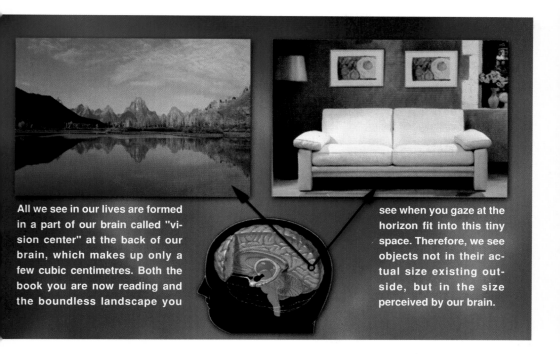

All we see in our lives are formed in a part of our brain called "vision center" at the back of our brain, which makes up only a few cubic centimetres. Both the book you are now reading and the boundless landscape you see when you gaze at the horizon fit into this tiny space. Therefore, we see objects not in their actual size existing outside, but in the size perceived by our brain.

you like is the interpretation of electrical signals by the brain. You can never reach the object in the outside; you can never see, smell or taste the chocolate itself. For instance, if taste nerves that travel to your brain are cut, nothing you eat at the moment will reach your brain; you will completely lose your sense of taste.

At this point, we come across with another fact: We can never be sure that what we feel when we taste a food and what another person feels when he tastes the same food, or what we perceive when we hear a voice and what another person perceives when he hears the same voice are the same. On this fact, Lincoln Barnett says that no one can know that another person perceives the colour red or hears the C note the same way as he himself does.[201]

Our sense of touch is no different than the others. When we touch an object, all information that will help us recognise the external world and objects are transmitted to the brain by the sense nerves on the skin. The feeling of touch is formed in our brain. Contrary to general belief, the place where we perceive the sense of touch is not at our finger tips or skin but at the centre of touch in our brain. As a result of the brain's assessment of electrical stimulations coming from objects to it, we feel different senses pertaining to those objects such as hardness or softness, or heat or cold. We derive all details that help us recognise an object from these stimulations. Concerning this important fact, the thoughts of two famous philosophers, B. Russell and L. J. J. Wittgeinstein are as follows;

> For instance, whether a lemon truly exists or not and how it came to exist cannot be questioned and investigated. A lemon consists merely of a taste sensed by the tongue, an odor sensed by the nose, a colour and shape sensed by the eye; and only these features of it can be subject to examination and assessment. Science can never know the physical world.[202]

It is impossible for us to reach the physical world. All objects around us are a collection of perceptions such as seeing, hearing, and touching. By processing the data in the centre of vision and in other sensory centres, our brain, throughout our lives, **confronts not the "original" of the matter existing outside us but rather the copy formed inside our brain.** It is at this point that we are misled by assuming that these copies are instances of real matter outside us.

"The External World" Inside Our Brain

As a result of the physical facts described so far, we may conclude the following. Everything we see, touch, hear, and perceive as matter", "the world" or "the universe" is nothing but electrical signals occurring in our brain.

Someone eating a fruit in fact confronts not the actual fruit but its perception in the brain. The object considered to be a "fruit" by the person actually consists of an electrical impression in the brain concerning the shape, taste, smell, and texture of the fruit. If the sight nerve travelling to the brain were to be severed suddenly, the image of the fruit would suddenly disappear. Or a disconnection in the nerve travelling from the sensors in the nose to the brain would completely interrupt the sense of smell. Simply put, the fruit is nothing but the interpretation of electrical signals by the brain.

Another point to be considered is **the sense of distance**. Distance, which is to say the distance between you and this book, is only a feeling of emptiness formed in your brain. Objects that seem to be distant in that person's view also exist in the brain. For instance, someone who watches the

As a result of artificial stimulations, a physical world as true and realistic as the real one can be formed in our brain without the existence of physical world. As a result of artificial stimulations, a person may think that he is driving in his car, while he is actually sitting in his home.

stars in the sky assumes that they are millions of light-years away from him. Yet what he "sees" are really the stars inside himself, in his centre of vision. While you read these lines, you are, in truth, not inside the room you assume you are in; on the contrary, the room is inside you. Your seeing your body makes you think that you are inside it. **However, you must remember that your body, too, is an image formed inside your brain**.

The same applies to all your other perceptions. For instance, when you think that you hear the sound of the television in the next room, you are actually experiencing the sound inside your brain. You can neither prove that a room exists next to yours, nor that a sound comes from the television in that room. Both the sound you think to be coming from meters away and the conversation of a person right near you are perceived in a few centimetre-square centre of hearing in your brain. Apart from this centre of perception, no concept such as right, left, front or behind exists. That is, sound does not come to you from the right, from the left or from the air; **there is no direction from which the sound comes**.

The smells that you perceive are like that too; none of them reach you from a long distance. You suppose that the end-effects formed in your centre of smell are the smell of the objects in the outside. However, just as the image of a rose is in your centre of vision, so the smell of this rose is in your centre of smell; you can never know whether the original of that rose or smell really exists outside.

The "external world" presented to us by our perceptions is merely a collection of the electrical signals reaching our brain. Throughout our lives, these signals are processed by our brain and we live without recognising that we are mistaken in assuming that these are the original versions of matter existing in the "external world". We are misled because we can never reach the matter itself by means of our senses.

Moreover it is again our brain that interprets and attributes meaning to the signals that we assume to be the "external world". For example, let us consider the sense of hearing. It is in fact our brain that transforms the sound waves in the "external world" into a symphony. That is to say, music is also a perception created by our brain. In the same manner, when we see colours, what reaches our eyes are merely electrical signals of **different wavelengths**. It is again our brain that transforms these signals into colours. **There are no colours in the "external world"**. Neither is the apple red nor is the sky blue nor the trees green. They are as they are just because we perceive them to be so. **The "external world" depends entirely on the perceiver.**

Even a slightest defect in the retina of the eye causes colour blindness. Some people perceive blue as green, some red as blue, and some all colours as different tones of grey. At this point, it does not matter whether the object outside is coloured or not.

The prominent thinker Berkeley also addresses this fact:

> At the beginning, it was believed that **colours, odours**, etc., "really exist", but subsequently such views were renounced, and it was seen that **they only exist in dependence on our sensations**.203

In conclusion, the reason we see objects coloured is not because they are coloured or

The findings of modern physics show that the universe is a collection of perceptions. The following question appears on the cover of the well-known American science magazine *New Scientist* which dealt with this fact in its 30 January 1999 issue: "Beyond Reality: Is the Universe Really a Frolic of Primal Information and Matter Just a Mirage?"

because they have an independent material existence outside ourselves. The truth of the matter is rather that **all the qualities we ascribe to objects are inside us and not in the "external world"**.

So what remains of the "external world"?

Is The Existence Of The "External World" Indispensable?

So far we have been speaking repeatedly of an "external world" and a world of perceptions formed in our brain, the latter of which is the one we see. However since we can never actually reach the "external world", how can we be sure that such a world really exists?

Actually we cannot. Since each object is only a collection of perceptions and those perceptions exist only in the mind, it is more accurate to say that **the only world that we deal with is the world of perceptions**. The only world we know of is the world that exists in our mind: the one that is designed, recorded, and made vivid there; the one, in short, that is created within our mind. This is the only world we can be sure of.

We can never prove that the perceptions we observe in our brain have material correlates. Those perceptions may well be coming from an "artificial" source.

It is possible to observe this. False stimulations can produce in our brain an entirely imaginary "material world". For example, let us think of a very developed recording instrument where all kinds of electrical signals can be recorded. First, let us transmit all the data related to a setting (including body image) to this instrument by transforming them into electrical signals. Second, let us imagine that you can have your brain survive apart from your body. Lastly, let us connect the recording instrument to the brain with electrodes that will function as nerves and send the pre-recorded data to the brain. In this state, you will feel yourself as if you are living in this artificially created setting. For instance, you can easily believe that you are driving fast on a highway. It never becomes possible to understand that you consist of nothing but your brain. This is because what is needed to form a world within your brain is not the existence of a real world but rather the availability of stimulations. It is perfectly possible that these stimulations could be coming from an artificial source, such as a recorder.

In that connection, distinguished science philosopher Bertrand Russell wrote;

As to the sense of touch when we press the table with our fingers, that is an electric disturbance on the electrons and protons of our fingertips, produced, according to modern physics, by the proximity of the electrons and protons in the table. **If the same disturbance in our finger-tips arose in any other way, we should have the sensations, in spite of there being no table.**[204]

It is indeed very easy for us to be deceived into deeming perceptions without any material correlates as real. We often experience this feeling in our dreams. In our dreams, we experience events, see people, objects and settings that seem completely real. However, they are all nothing but mere perceptions. There is no basic difference between the dream and the "real world"; both of them are experienced in the brain.

Who Is The Perceiver?

As we have related so far, there is no doubt of the fact that the world we think we are inhabiting and that we call the "external world" is created inside our brain. However, here arises the question of primary importance. If all the physical events that we know of are intrinsically perceptions, what about our brain? Since our brain is a part of the physical world just like our arm, leg, or any other object, it also should be a perception just like all other objects.

An example about dreams will illuminate the subject further. Let us think that we see the dream within our brain in accordance with what has been said so far. In the dream, we will have an imaginary body, an imaginary arm, an imaginary eye, and an imaginary brain. If during our dream we were asked "where do you see?", we would answer "I see in my brain". Yet, actually there is not any brain to talk about, but an imaginary head and an imaginary brain. The seer of the images is not the imaginary brain in the dream, but a "being" that is far "superior" to it.

We know that there is no physical distinction between the setting of a dream and the setting we call real life. So when we are asked in the setting we call real life the above question of "where do you see", it would be just as meaningless to answer "in my brain" as in the example above. In both conditions, the entity that sees and perceives is not the brain, which is after all only a hunk of meat.

When the brain is analysed, it is seen that there is nothing in it but lipid and protein molecules, which also exist in other living organisms. This means that within the piece of meat we call our "brain", there is nothing to observe the images, to constitute consciousness, or to create the being we call "myself".

R.L. Gregory refers to a mistake people make in relation to the perception of images in the brain:

> There is a temptation, which must be avoided, to say that the eyes produce pictures in the brain. A picture in the brain suggests the need of some kind of internal eye to see it - but this would need a further eye to see its picture... and so on in an endless regress of eyes and pictures. This is absurd.[205]

This is the very point which puts the materialists, who do not hold anything but the matter as true, in a quandary. To whom belongs "the eye inside" that sees, that perceives what it sees and reacts?

Karl Pribram also focused on this important question in the world of science and philosophy about who the perceiver is:

> Philosophers since the Greeks have speculated about the "ghost" in the machine, the "little man inside the little man" and so on. Where is the I -the entity that uses the brain? Who does the actual knowing? Or, as Saint Francis of Assisi once put it, "What we are looking for is what is looking."[206]

Now, think of this: The book in your hand, the room you are in, in brief, all the images in front of you are seen inside your brain. Is it the atoms that see these images? Blind, deaf, unconscious atoms? Why did some atoms acquire this quality whereas some did not? Do our acts of thinking, comprehending, remembering, being delighted, being unhappy, and everything else consist of the electrochemical reactions between these atoms?

When we ponder these questions, we see that there is no sense in looking for will in atoms. It is clear that the being who sees, hears, and feels is a supra-material being. This being is "alive" and it is neither matter nor an image of matter. This being associates with the perceptions in front of it by using the image of our body.

This being is the "soul".

The aggregate of perceptions we call the "material world" is a dream observed by this soul. Just as the body we possess and the material world we see in our dreams have no reality, the universe we occupy and the body we possess also have no material reality.

The real being is the soul. Matter consists merely of perceptions

 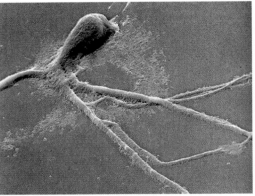

The brain is a heap of cells made up of protein and fat molecules. It is formed of nerve cells called neurons. There is no power in this piece of meat to observe the images, to constitute consciousness, or to create the being we call "myself".

viewed by the soul. The intelligent being that writes and reads these lines is not a heap of atoms and molecules-and the chemical reactions between them-but a "soul".

The Real Absolute Being

All these facts bring us face to face with a very significant question. If the thing we acknowledge to be the material world is merely comprised of perceptions seen by our soul, then what is the source of these perceptions?

In answering this question, we have to take the following fact into consideration: matter does not have a self-governing existence by itself. Since matter is a perception, it is something "artificial". That is, this perception must have been caused by another power, which means that it must in fact have been created. Moreover, this creation should be continuous. If there was not a continuous and consistent creation, then what we call matter would disappear and be lost. This may be likened to a television on which a picture is displayed as long as the signal continues to be broadcast. So, who makes our soul watch the stars, the earth, the plants, the people, our body and all else that we see?

It is very evident that there exists a supreme Creator, Who has created the entire material universe, that is, the sum of perceptions, and Who continues His creation ceaselessly. Since this Creator displays such a magnificent creation, he surely has eternal power and might. This Creator describes us Himself, the universe and the reason of our existence through the book He has sent down.

This Creator is God and the name of His Book is the Qur'an.

The facts that the heavens and the earth, that is, the universe is not stable, that their presence is only made possible by God's creation and that they will disappear when He ends this creation, are all explained in a verse as follows:

> It is God Who sustains the heavens and the earth, lest they cease (to function): and if they should fail, there is none -not one- can sustain them thereafter: Verily He is Most Forbearing, Oft-Forgiving. (Surah Fatir, 41)

As we mentioned at the beginning, some people have no genuine understanding of God and so they imagine Him as a being present somewhere in the heavens and not really intervening in worldly affairs. The basis of this logic actually lies in the thought that the universe is an assembly of matter and God is "outside" this material world, in a far away place. In some false religions, belief in God is limited to this understanding.

However, as we have considered so far, matter is composed only of sensations. And the only real absolute being is God. That means that it is **only God that exists: everything except Him are images**. Consequently, it is impossible to conceive God as a separate being outside this whole mass of matter. **God is surely "everywhere" and encompasses all.** This reality is explained in the Qur'an as follows;

> God! There is no god but He, the Living, the Self-subsisting, Eternal. No slumber can seize Him nor sleep. His are all things in the heavens and on earth. Who is there can intercede in His presence except as He permits? He knows what (appears to His creatures as) before or after or behind them. Nor shall they compass aught of His knowledge except as He wills. **His Throne extends over the heavens and the earth**, and He feels no fatigue in guarding and preserving them for He is the Most High, the Supreme (in glory). (Surat al-Baqara, 255)

The fact that God is not bound with space and that He encompasses everything roundabout is stated in another verse as follows:

> To God belong the east and the west: **Whithersoever you turn, there is the presence of God.** For God is all-Pervading, all-Knowing. (Surat al-Baqara, 115)

Since material beings are each a perception, they cannot see God; but God sees the matter He created in all its forms. In the Qur'an, this fact is stated thus: "**No vision can grasp Him, but His grasp is over all vision**" (Surat al-Anaam, 103)

That is, we cannot perceive God with our eyes, but God has thoroughly encompassed our inside, outside, looks and thoughts. We cannot utter any word but with His knowledge, nor can we even take a breath.

While we watch these sensory perceptions in the course of our lives, the closest being to us is not any one of these sensations, but God Himself. The secret of the following verse in the Qur'an is concealed in this reality: "It was We Who created man, and We know what dark suggestions his soul makes to him: for **We are nearer to him than (his) jugular vein.**" (Surah Qaf: 16) When a person thinks that his body is made up of "matter", he cannot comprehend this important fact. If he takes his brain to be "himself", then the place he accepts to be the outside will be 20-30 cms away from him. However, when he conceives that everything he knows as matter, is imagination, notions such as outside, inside, or near lose meaning. **God has encompassed him and He is "infinitely close" to him**.

> Why is it not then that when it (soul) comes up to the throat, and you at that time look on, We are nearer to him than you, but you see not. (Surat al-Waqia, 83-85)

God informs men that He is "**infinitely close**" to them with the verse "When My servants ask you concerning Me, **I am indeed close (to them)**" (Surat al-Baqara, 186). Another verse relates the same fact: "We told you that **your Lord encompasses mankind round about.**" (Surat al-Isra, 60).

Man is misled by thinking that the being that is closest to him is himself. God, in truth, is closer to us even more than ourselves. He has called our attention to this point in the verse "Why is it not then that when it (soul) comes up to the throat, and you at that time look on, **We are nearer to him than you, but you see not.**" (Surat al-Waqia, 83-85). As informed in the verse, people live unaware of this phenomenal fact because they do not see it with their eyes.

On the other hand, it is impossible for man, who is nothing but an image, to have a power and will independent of God. The verse "But **God has created you and your handwork!**" (Surat as-Saaffat, 96) shows that everything we experience takes place under God's control. In the Qur'an, this reality is stated in the verse "**When you threw, it was not your act, but God's.**" (Surat al-Anfal, 17) whereby it is emphasised that no act is independent of God. Since a human being is an image, it cannot be itself which

If one ponders deeply on all that is said here, he will soon realise this amazing, extraordinary situation by himself: that all the events in the world are but mere imagination...

performs the act of throwing. However, God gives this image the feeling of the self. In reality, it is God Who performs all acts. So, if one takes the acts he does as his own, he evidently means to deceive himself.

This is the reality. A person may not want to concede this and may think of himself as a being independent of God; but this does not change a thing. Of course his unwise denial is again within God's will and wish.

Everything That You Possess Is Intrinsically Illusory

As it may be seen clearly, it is a scientific and logical fact that the "external world" has no materialistic reality and that it is a collection of images God perpetually presents to our soul. Nevertheless, people usually do not include, or rather do not want to include, everything in the concept of the "external world".

If you think on this issue sincerely and boldly, you come to realise that your house, your furniture in it, your car-perhaps recently bought, your office, your jewels, your bank account, your wardrobe, your spouse, your children, your colleagues, and all else that you possess are in fact included in this imaginary external world projected to you. Everything you see, hear, or smell-in short-perceive with your five senses around you is a part of this "imaginary world" the voice of your favourite singer, the hardness of the chair you sit on, a perfume whose smell you like, the sun that keeps you warm, a flower with beautiful colours, a bird flying in front of your window, a speedboat moving swiftly on the water, your fertile garden, the computer you use at your job, or your hi-fi that has the most advanced technology in the world...

This is the reality, because the world is only a collection of images created to test man. People are tested all through their limited lives with perceptions bearing no reality. These perceptions are intentionally presented as appealing and attractive. This fact is mentioned in the Qur'an:

> Fair in the eyes of men is the love of things they covet: Women and sons; Heaped-up hoards of gold and silver; horses branded (for blood and excellence); and (wealth of) cattle and well-tilled land. Such are the possessions of this world's life; but in nearness to God is the best of the goals (to return to). (Surat Aal-e Imran, 14)

Most people cast their religion away for the lure of property, wealth, heaped-up hoards of gold and silver, dollars, jewels, bank accounts, credit cards, wardrobe-full clothes, late-model cars, in short, all forms of prosperity they either possess or strive to possess and they concentrate only on this world while forgetting the hereafter. They are deceived by the "fair and alluring" face of the life of this world, and fail to keep up prayer, give charity to the poor, and perform worship that will make them prosper in the hereafter by saying "I have things to do", "I have ideals", "I have responsibilities", "I do not have enough time", "I have things to complete", "I will do them in the future". They consume their lives by trying to prosper only in this world. In the verse, **"They know but the outer (things) in the life of this world: but of the End of things they are heedless."** (Surat ar-Room, 7), this misconception is described.

The fact we describe in this chapter, namely the fact that everything is an image, is very important for its implication that it renders all the lusts and boundaries meaningless. The verification of this fact makes it clear that everything people possess and toil to possess, their wealth made with greed, their children with whom they boast, their spouses who they consider to be closest to them, their friends, their dearest bodies, their rank which they hold to be a superiority, the schools they have attended, the holidays they have been are nothing but mere illusion. Therefore, all the efforts put, the time spent, and the greed felt prove to be in unavailing.

This is why some people unwittingly make fools of themselves when they boast of their wealth and properties or of their "yachts, helicopters, factories, holdings, manors and lands" as if they ever really existed. Those well-to-do people who ostentatiously saunter up and down in their yachts, show off with their cars, keep talking about their wealth, suppose that their post rank them higher than everyone else and keep thinking that they are successful because of all this, should actually think what kind of a state they would find themselves in once they realise that their success is nothing but an illusion.

In fact, these scenes are many times seen in dreams as well. In their dreams, they also have houses, fast cars, extremely precious jewels, rolls of dollars, and loads of gold and silver. In their dreams, they are also positioned in a high rank, own factories with thousands of workers, possess power to rule over many people, put on clothes that make every one admire them... Just as boasting about one's possessions in one's dream causes a person to be ridiculed, he is sure to be equally ridiculed for boasting of images he sees in this world. After all, both what he sees in his dreams and what he relates to in this world are mere images in his mind.

Similarly the way people react to the events they experience in the world is to make them feel ashamed when they realise the reality. Those who fiercely fight with each other, those who rave furiously, who swindle, who take bribes, who commit forgery, who lie, who covetously withhold their money, who do wrong to people, who beat and curse others, raging aggressors, those who are full of passion for office and rank, who practice envy, who try to show off, who try to sanctify themselves and all others will be disgraced when they realise that they have committed all of these deeds in a dream.

Since it is God Who creates all these images, the Ultimate Owner of everything is God alone. This fact is stressed in the Qur'an:

> **But to God belong all things in the heavens and on earth: And He it is that Encompasses all things.** (Surat an-Nisa, 126)

It is a great foolishness to cast religion away at the cost of imaginary passions and thus lose the eternal life.

At this stage, one point should be well grasped: it is not said here that the fact you face predicates that "all the possessions, wealth, children, spouses, friends, rank you have with which you are being stingy will vanish sooner or later, and therefore they do not have any meaning". It is rather said that "all the possessions you seem to have in fact do not exist at all, but they are merely a dream and composed of images God shows to test you". As you see, there is a big difference between the two statements.

Although one does not want to acknowledge this fact right away and would rather deceive himself by assuming everything he has truly exists, he is finally to die and in the hereafter everything is to become clear when he is recreated. On that day "**sharp is one's sight**" (Surah Qaf, 22) and he is apt to see everything much more clearly. However, if he has spent his life chasing after imaginary aims, he is going to wish he had never lived his life and say "Ah! Would that (Death) had made an end of me! Of no profit to me has been my wealth! My power has perished from me!" (Surat al-Haqqaa, 27-29)

What a wise man should do, on the other hand, is to try to understand the greatest reality of the universe here on this world, while he still has time. Otherwise, he is to spend all his life running after dreams and face a grievous penalty in the end. In the Qur'an, the final state of those people who run after illusions (or mirages) on this world and forget their Creator, is stated as follows;

> **But the Unbelievers, their deeds are like a mirage in sandy deserts,** which the man parched with thirst mistakes for water; until when he comes up to it, he finds it to be nothing: But he finds God (ever) with him, and God will pay him his account: and God is swift in taking account. (Surat an-Noor, 39)

Logical Deficiencies Of The Materialists

Since the beginning of this chapter, it is clearly stated that matter is not an absolute being as the materialists claim but rather a collection of senses God creates. Materialists resist in an extremely dogmatic manner this evident reality which destroys their philosophy and bring forward baseless anti-theses.

For example, one of the biggest advocates of the materialist philosophy in the 20th century, an ardent Marxist, **George Politzer**, gave the "**bus example**" as the "greatest evidence" for the existence of matter. According to Politzer, philosophers who think that matter is a perception also run away when they see a bus and this is the proof of the physical existence of matter.[207]

> But the Unbelievers, their deeds are like a mirage in sandy deserts, which the man parched with thirst mistakes for water; until when he comes up to it, he finds it to be nothing: But he finds God (ever) with him, and God will pay him his account: and God is swift in taking account.
> (Surat an-Noor, 39)

When another famous materialist, Johnson, was told that matter is a collection of perceptions, he tried to "prove" the physical existence of stones by giving them a kick.[208]

A similar example is given by **Friedrich Engels**, the mentor of Politzer and the founder of dialectic materialism along with Marx, who wrote "**if the cakes we eat were mere perceptions, they would not stop our hunger**".[209]

There are similar examples and impetuous sentences such as "**you understand the existence of matter when you are slapped in the face**" in the books of famous materialists such as **Marx**, **Engels**, **Lenin**, and others.

The disorder in comprehension that gives way to these examples of the materialists is their interpreting the explanation of "matter is a perception" as "matter is a trick of light". They think that the concept of perception is only limited to sight and that perceptions like touching have a physical correlate. A bus knocking a man down makes them say "Look, it crashed, therefore it is not a perception". What they do not understand is that all perceptions experienced during a bus crash such as hardness, collision, and pain are formed in the brain.

The Example Of Dreams

The best example to explain this reality are dreams. A person can experience very realistic events in his dream. He can roll down the stairs and break his leg, have a serious car accident, get stuck under a bus, or eat a cake and be satiated. Similar events to those experienced in our daily lives are also experienced in dreams with the same persuasiveness and rousing the same feelings in us.

A person who dreams that he is knocked down by a bus can open his eyes in a hospital again in his dream and understand that he is disabled, but this all would be a dream. He can also dream that he dies in a car crash, angels of death take his soul, and his life in the hereafter begins. (This event is experienced in the same manner in this life, which is a perception just like the dream.)

This person very sharply perceives the images, sounds, feeling of hardness, light, colours, and all other feelings pertaining to the event he experiences in his dream. The perceptions he perceives in his dream are as natural as the ones in "real" life. The cake he eats in his dream satiates him although it is a mere perception, because being satiated is also a perception. However, in reality, this person is lying in his bed at that moment. There are no stairs, no traffic, no buses to consider. The dreaming person experiences and sees perceptions and feelings that do not exist in the external world. The fact that in our dreams, we experience, see, and feel events with no physical correlates in the "external world" very clearly reveals that the "external world" absolutely consists of mere perceptions.

Those who believe in the materialist philosophy, and particularly the Marxists, are enraged when they are told about this reality, the essence of matter. They quote examples from the superficial reasoning of **Marx**, **Engels**, or **Lenin** and make emotional declarations.

However, these persons must think that they can also make these declarations in their dreams. In their dream, they can also read "Das Kapital", participate in meetings, fight with the police, get hit on the head, and moreover, feel the pain of their wounds. When they are asked in their dreams, they will think that what they experience in their dreams also consists of "absolute matter"-just as they assume the things they see when they are awake are "absolute matter". However, be it in their dream or in their daily lives, all that they see, experience, or feel consists only of perceptions.

THE WORLD IN DREAMS

For you, reality is all that can be touched with the hand and seen with the eye. In your dreams you can also "touch with your hand and see with your eye", but in reality, you have neither hand nor eye, nor is there anything that can be touched or seen. There is no material reality that makes these things happen except your brain. You are simply being deceived.

What is it that separates real life and the dreams from one another? Ultimately, both forms of life are brought into being within the brain. If we are able to live easily in an unreal world during our dreams, the same thing can equally be true for the world we live in. When we wake up from a dream, there is no logical reason for not thinking that we have entered a longer dream that we call "real life". The reason we consider our dream to be fancy and the world as real is nothing but a product of our habits and prejudices. This suggests that we may well be awoken from the life on earth which we think we are living right now, just as we are awoken from a dream.

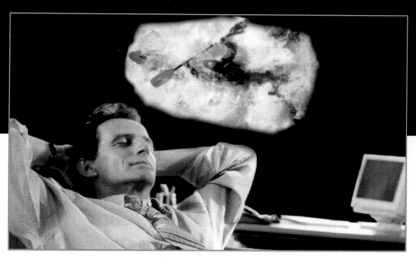

The Example Of Connecting The Nerves In Parallel

Let us consider the car crash example of Politzer: In this accident, if the crushed person's nerves travelling from his five senses to his brain, were connected to another person's, for instance Politzer's brain, with a parallel connection, at the moment the bus hit that person, it would also hit Politzer, who is sitting at his home at that moment. Better to say, all the feelings experienced by that person having the accident would be experienced by Politzer, just like the same song is listened from two different loudspeakers connected to the same tape recorder. Politzer will feel, see, and experience the braking sound of the bus, the touch of the bus on his body, the images of a broken arm and shedding blood, fracture aches, the images of his entering the operation room, the hardness of the plaster cast, and the feebleness of his arm.

Every other person connected to the man's nerves in parallel would experience the accident from beginning to end just like Politzer. If the man in the accident fell into a coma, they would all fall into a coma. Moreover, if all the perceptions pertaining to the car accident were recorded in a device and if all these perceptions were transmitted to a person, the bus would knock this person down many times.

So, which one of the buses hitting those people is real? The materialist philosophy has no consistent answer to this question. The right answer is that they all experience the car accident in all its details in their own minds.

The same principle applies to the cake and stone examples. If the nerves of the sense organs of Engels, who felt the satiety and fullness of the cake in his stomach after eating a cake, were connected to a second person's brain in parallel, that person would also feel full when Engels ate the cake and was satiated. If the nerves of Johnson, who felt pain in his foot when he delivered a sound kick to a stone, were connected to a second person in parallel, that person would feel the same pain.

So, which cake or which stone is the real one? The materialist philosophy again falls short of giving a consistent answer to this question. The correct and consistent answer is this: both Engels and the second person have eaten the cake in their minds and are satiated; both Johnson and the second person have fully experienced the moment of striking the stone in their minds.

Let us make a change in the example we gave about Politzer: let us connect the nerves of the man hit by the bus to Politzer's brain, and the nerves of Politzer sitting in his house to that man's brain, who is hit by the bus. In this case, Politzer will think that a bus has hit him although when he is sitting in his house; and the man actually hit by the bus will never feel the impact of the accident and think that he is sitting in Politzer's house. The very same logic may be applied to the cake and the stone examples.

As is to be seen, it is not possible for man to transcend his senses and break free of them. In this respect, a man's soul can be subjected to all kinds of representations although it has no physical body and no material existence and lacks material weight. It is not possible for a person to realise this because he assumes these three-dimensional images to be real and is absolutely certain of their existence because everybody depends on perceptions that are caused to be felt by his sensory organs.

The famous British philosopher David Hume expresses his thoughts on this fact:

> Frankly speaking, when I include myself in what I call "myself", I always come across with a specific perception pertaining to hot or cold, light or shadow, love or hatred, sour or sweet or some other notion. Without the existence of a perception, I can never capture myself in a particular time and **I can observe nothing but perception**.[210]

The Formation Of Perceptions In The Brain Is Not Philosophy But Scientific Fact

Materialists claim that what we have been saying here is a philosophical view. However, to hold that the "external world", as we call it, is a collection of perceptions is not a matter of philosophy but a plain scientific fact. How the image and feelings form in the brain is taught in all medical schools in detail. These facts, proven by the 20th-century science, and particularly by physics, clearly show that matter does not have an absolute reality and that everyone in a sense is watching the "monitor in his brain".

Everyone who believes in science, be he an atheist, Buddhist, or anyone who holds another view has to accept this fact. A materialist might deny the existence of a Creator yet he cannot deny this scientific reality.

The inability of Karl Marx, Friedrich Engels, Georges Politzer and others in comprehending such a simple and evident fact is still startling although the level of scientific understanding and possibilities of their times were insufficient. In our time, science and technology are highly advanced and recent discoveries make it easier to comprehend this fact. Materialists, on the other hand, are flooded with the fear of both comprehending this fact, though partially, and realising how definitely it demolishes their philosophy.

The Great Fear Of The Materialists

For a while, no substantial backlash came from the Turkish materialist circles against the subject brought up in this book, that is, the fact that matter is a mere perception. This had given us the impression that our point was not made so clear and that it needed further explanation. Yet before long, it was revealed that materialists felt quite uneasy about the popularity of this subject and moreover, felt a great fear all about this.

For a while, materialists have been loudly pronouncing their fear and panic in their publications, conferences and panels. Their agitated and hopeless discourse implies that they are suffering from a severe intellectual crisis. The scientific collapse of the theory of evolution, the so-called basis of their philosophy, had already come as a great shock to them. Now, they come to realise that they start to lose the matter itself, which is a greater mainstay for them than Darwinism, and they experience an even greater shock. They declare that this issue is the "biggest threat" for them and that it totally "demolishes their cultural fabric".

One of those who expressed this anxiety and panic felt by the materialist circles in the most outspoken way was Renan Pekunlu, an academician as well as a writer of *Bilim ve Utopya* (Science and Utopia) periodical which has assumed the task of defending materialism. Both in his articles in *Bilim ve Utopya* and in the panels he attended, Pekunlu presented the book *Evolution Deceit* as the number one "threat" to materialism. What disturbed Pekunlu even more than the chapters that invalidated Darwinism was the part you are currently reading. To his readers and (only a handful of) audience, Pekunlu delivered the message "do not let yourselves be carried away by the indoctrination of idealism and keep your faith in materialism" and showed Vladimir I. Lenin, the leader of the bloody communist revolution in Russia, as reference. Advising everyone to read Lenin's century-old book titled *Materialism and Empirio-Criticism,* all Pekunlu did was to repeat the counsels of Lenin stating "do not think over this issue, or you will lose track of materialism and be carried away by religion". In an article he wrote in the aforementioned periodical, he quoted the following lines from Lenin:

> Once you deny objective reality, given us in sensation, you have already lost every weapon against fideism, for you have slipped into agnosticism or subjectivism-and that is all that fideism requires. **A single claw ensnared, and the bird is lost.** And our Machists have all become ensnared in idealism, that is, in a diluted, subtle fideism; they became ensnared from the moment they took "sensation" not as an image of the external world but as a special "element". It is nobody's sensation, nobody's mind, nobody's spirit, nobody's will.[211]

These words explicitly demonstrate that the fact which Lenin alarmingly realised and wanted to take out both from his mind and the minds of his "comrades" also disturbs contemporary materialists in a similar way. However, Pekunlu and other materialists suffer a yet greater distress; because they are aware that this fact is now being put forward in a far more explicit, certain and convincing way than 100 years ago. For the first time in world history, this subject is being explained in such an irresistible way,.

Nevertheless, the general picture is that a great number of materialist scientists still take a very superficial stand against the fact that "matter is nothing but an illusion". The subject explained in this chapter is **one of the most important and most exciting subjects** that one can ever come across in one's life. It is rather unlikely that they would have faced such a crucial subject before. Still, the reactions of these scientists or the manner they employ in their speeches and articles hint how shallow and superficial their comprehension is.

It is so much so that the reactions of some materialists to the subject discussed here show that their blind adherence to materialism has caused some kind of a harm in their logic and for this reason, they are far removed from comprehending the subject. For instance Alaattin Senel, also an academician and a writer for *Bilim ve Utopya,* gave similar messages as Rennan Pekunlu saying "**Forget the collapse of Darwinism, the real threatening subject is this one**", and made demands such as "so you prove what you tell" sensing that his own philosophy has no basis. What is more interesting is that this writer himself has written lines revealing that he can by no means grasp this fact which he considers to be a menace.

For instance, in an article where he exclusively discussed this subject, Senel accepts that the external world is perceived in the brain as an image. However, he then goes on to claim that images are divided into two as those having physical correlates and those that do not, and that images pertaining to the external world have physical correlates. In order to support his assertion, he gives "the example of telephone". In sum-

Turkish materialist writer Rennan Pekunlu says that "the theory of evolution is not so important, the real threat is this subject", because he is aware that this subject nullifies matter, the only concept he has faith in.

mary, he wrote: "I do not know whether the images in my brain have correlates in the external world or not, but the same thing applies when I speak on the phone. When I speak on the telephone, I cannot see the person I am speaking to but I can have this conversation confirmed when I later see him face to face."[212]

By saying so, this writer actually means the following: "If we doubt our perceptions, we can look at the matter itself and check its reality." However, this is an evident misconception because it is impossible for us to reach the matter itself. **We can never get out of our mind and know what is "outside"**. Whether the voice on the phone has a correlate or not can be confirmed by the person on the phone. However, this confirmation is also imagery experienced by the mind.

As a matter of fact, these people also experience the same events in their dreams. For instance, Senel may also see in his dream that he speaks on the phone and then have this conversation confirmed by the person he spoke to. Or, Pekunlu may in his dream feel as facing "a serious threat" and advise people to read century-old books of Lenin. However, no matter what they do, these materialists can never deny the fact that the events they have experienced and the people they have talked to in their dreams were nothing but perceptions.

By whom, then, will one confirm whether the images in the brain have correlates or not? By the images in his brain again? Without doubt, it is impossible for materialists to find a source of information that can yield data concerning the outside of the brain and confirm it.

Conceding that all perceptions are formed in the brain but assuming that one can step "out" of this and have the perceptions confirmed by the real external world reveals that the perceptive capacity of the person is limited and that he has a distorted reasoning.

However, the fact told here can easily be captured by a person with a normal level of understanding and reasoning. Each unbiased person would know, in relation to all that we have said, that it is not possible for him to test the existence of the external world with his senses. Yet, it appears that blind adherence to materialism distorts the reasoning capability of people. For this reason, contemporary materialists display severe logical flaws just like their mentors who tried to "prove" the existence of matter by kicking stones or eating cakes.

It also has to be stated that this is not an astonishing situation; because, inability to understand is a common trait of all unbelievers. In the Qur'an, God particularly states that they are "**a people without understanding**" (Surat al-Baqara 171)

Materialists Have Fallen Into The Biggest Trap In History

The atmosphere of panic sweeping through the materialist circles in Turkey of which we have mentioned only a few examples here shows that materialists face an utter defeat such as they have never met in history. The fact that matter is simply a perception has been proven by modern science and it is put forward in a very clear, straightforward and forceful way. It only remains for the materialists to see the collapse of the entire material world they blindly believe and rely on.

Throughout the history of humanity, materialist thought always existed. Being very assured of themselves and the philosophy they believe in, materialists revolted against God who has created them. The scenario they formulated maintained that matter had no beginning or end, and that none of it could possibly have a Creator. While they denied God just because of their arrogance, they took refuge in matter which they held to have a real existence. They were so confident of this philosophy that they thought it would never be possible to put forth an argument disproving it.

That is why the facts told in this book regarding the real nature of matter surprised these people so much. What has been told here destroyed the very basis of their philosophy and left no ground for further discussion. Matter, upon which they based all their thoughts, lives, arrogance and denial, vanished all of a sudden. **How can materialism exist when matter does not?**

One of the attributes of God is His plotting against the unbelievers. This is stated in the verse "They plot and plan, and God too plans; but **the best of planners is God.**" (Surat al- Anfal, 30)

God entrapped materialists by making them assume that matter exists and so doing, humiliated them in an unseen way. Materialists deemed their possessions, status, rank, the society they belong, the whole world and everything else to be existing and moreover, grew arrogant against God by relying on these. They revolted against God by being boastful and added to their unbelief. While so doing, they totally relied on matter. Yet, they are so lacking in understanding that they fail to think that God compasses them round about. God announces the state to which the unbelievers are led as a result of their thick-headedness:

> Or do they intend a plot (against you)? But **those who defy God are themselves involved in a Plot!** (Surat At- Tur, 42)

This is most probably the biggest defeat in history. While growing arrogant of their own accord, materialists have been tricked and suffered a serious defeat in the war they waged against God by bringing up something monstrous against Him. The verse **"Thus have We placed leaders in every town, its wicked men, to plot (and burrow) therein: but they only plot against their own souls, and they perceive it not"** announces how unconscious these people who revolt against their Creator are, and how they will end up (Surat al- Anaam, 123). In another verse the same fact is related as:

> Fain would they deceive God and those who believe, but **they only deceive themselves, and realise (it) not!** (Surat al- Baqara, 9)

While the unbelievers try to plot, they do not realise a very important fact as stressed by the words "they only deceive themselves, and realise (it) not!" in the verse. This is the fact that everything they experience is an imagery designed to be perceived by them, and all plots they devise are simply images formed in their brain just like every other act they perform. Their folly has made them forget that they are all alone with God and, hence, they are entrapped in their own devious plans.

No less than those unbelievers who lived in the past, those living today too face a reality that will shatter their devious plans from its basis. With the verse **"...feeble indeed is the cunning of Satan"** (Surat An-Nisa, 76), God has stated that these plots were doomed to end with failure the day they were hatched, and gave the good tidings to believers with the verse **"...not the least harm will their cunning do to you"**. (Surat Aal-E-Imran, 120)

In another verse God states: **"But the Unbelievers,- their deeds are like a mirage in sandy deserts, which the man parched with thirst mistakes for water; until when he comes up to it, he finds it to be nothing"** (Surat an- Noor 39). Materialism, too, becomes a "mirage" for the rebellious just like it is stated in this verse; when they have recourse to it, they find it to be nothing but an illusion. God has deceived them with such a mirage, and beguiled them into perceiving this whole collection of images as real. All those "eminent" people, professors, astronomers, biologists, physicists, and all others regardless of their rank and post are simply deceived like children, and are humiliated because they took matter as their god. Assuming a collection of images to be absolute, they based their philosophy and ideology on it, got involved in serious discussions, and adopted a so-called "intellectual" discourse. They deemed themselves to be wise enough to offer an argument about the truth of the universe and, more importantly, dispute about God with their limited intelligence. God explains their situation in the following verse:

> **And (the unbelievers) plotted and planned, and God too planned,** and the best of planners is God. (Surat Aal-E-Imran 54)

It may be possible to escape from some plots; however, this plan of God against the unbelievers is so firm that there is no way of escape from it. No matter what they do or to whom they appeal, they can never find a helper other than God. As God informs in the Qur'an, "**they shall not find for them other than God a patron or a help.**" (Surat an-Nisa, 173)

Materialists never expected to fall into such a trap. Having all the means of the 20th centry at their disposal, they thought that they could grow obstinate in their denial and drag people to disbelief. This ever-lasting mentality of unbelievers and their end are described as follows in the Qur'an:

> **They plotted and planned,** but **We too planned, even while they perceived it not.** Then see what was the end of their plot!- this, that **We destroyed them and their people, all (of them).** (Surat an- Naml 50-51)

This, in another sense, is what the fact stated in the verses comes to mean: materialists are made to realise that everything they own is but an illusion, and therefore **everything they possess has been destroyed**. As they witness their possessions, factories, gold, dollars, children, spouses, friends, rank and status, and even their own bodies, all of which they deem to exist, slipping away from their hands, they are "**destroyed**" in the words of the 51st verse of Surat an-Naml. At this point, they are no more matter but souls.

No doubt, realising this truth is the worst possible thing for the materialists. The fact that everything they possess is but an illusion, is tantamount, in their own words, to "death before dying" in this world.

This fact leaves them all alone with God. With the verse, "**Leave Me alone, (to deal) with the (creature) whom I created (bare and) alone!**", God has called us to attention that each human being is, in truth, all alone in His presence. (Surat Al-Muddaththir, 11) This remarkable fact is repeated in many other verses:

> "And behold! you come to us **bare and alone** as We created you for the first time: you have left behind you all (the favors) which We bestowed on you..." (Surat al-Anaam, 94)

> And each one of them will come unto Him on the Day of Resurrection, **alone**. (Surat Maryam, 95)

This, in another sense, is what the fact stated in the verses comes to mean: Those who take matter as their god have come from God and returned to Him. They have submitted their wills to God whether they want it or not. Now they wait for the Day of Judgment on which every one of them will be called to account. Though however unwilling they may be to understand it...

Conclusion

The subject we have explained so far is one of the greatest truths that will ever be told to you in your lifetime. Proving that the whole material world is in reality a "**image**", this subject is the key to comprehending the existence of God and His creations and of understanding that He is the only absolute being.

The person who understands this subject realises that the world is not the sort of place it is surmised to be by most people. The world is not an absolute place with a true existence as supposed by those who wander aimless about the streets, who get into fights in pubs, who show off in luxurious cafes, who brag about their property, or who dedicate their lives to hollow aims. The world is only a collection of perceptions, an illusion. All of the people we have cited above are only images who watch these perceptions in their minds: yet they are not aware of this.

This concept is very important for it undermines the **materialist philosophy** that denies the existence of God and causes it to collapse. This is the reason why materialists like **Marx**, **Engels**, and **Lenin** felt panic, became enraged, and warned their followers "not to think over" this concept when they were told about it. As a matter of fact, such people are in such a state of mental deficiency that they cannot even comprehend the fact that perceptions are formed inside the brain. They assume that the world they watch in their brain is the "external world" and they cannot comprehend the obvious evidence to the contrary.

This unawareness is the outcome of the lack of wisdom God gives to disbelievers. As it is said in the Qur'an, the unbelievers "**have hearts wherewith they understand not**, eyes wherewith they see not, and ears wherewith they hear not. They are like cattle-nay more misguided: for they are heedless (of warning)." (Surat al-Araf, 179)

You can explore beyond this point by using the power of your personal reflection. For this, you have to concentrate, devote your attention, and ponder on the way you see the objects around you and the way you feel their touch. If you think heedfully, you can feel that the wise being that sees, hears, touches, thinks, and reads this book at this moment is only a soul and watches the perceptions called "matter" on a screen. The person who comprehends this is considered to have moved away from the domain of the material world that deceives a major part of humanity and to have entered the domain of true existence.

This reality has been understood by a number of theists or philosophers throughout history. Islamic intellectuals such as Imam Rabbani, Muhyiddin Ibn Arabi and Mevlana Cami realised this fact from the signs of the Qur'an and by using their reason. Some Western philosophers like George Berkeley have grasped the same reality through reason. Imam Rabbani wrote in his Mektubat (Letters) that the whole material universe is an "illusion and supposition (perception)" and that the only absolute being is God:

> God... The substance of these beings which He created is but nothingness... He created all at **the sphere of senses and illusions...** The existence of the universe is at the sphere of senses and illusions, and it is not material... In real, there is nothing in the outside except the Glorious Being, (who is God).[213]

Imam Rabbani explicitly stated that all images presented to man are but an illusion, and that they have no originals in the "outside".

> This imaginary cycle is portrayed in imagination. It is seen to the extent that it is portrayed. Yet **with the mind's eye**. In the outside, it seems as if it is seen with the head's eye. However, the case is not so. It has neither a designation nor a trace in the outside. There is no circumstance to be seen. Even the face

of a person reflecting on a mirror is like that. It has no constancy in the outside. No doubt, both its constancy and image are in the **IMAGINATION**. God is He Who knows Best.214

Mevlana Cami stated the same fact which he discovered following the signs of the Qur'an and by using his wit: "**Whatever there is in the universe are senses and illusions. They are either like reflections in mirrors or shadows**".

However, the number of those who have understood this fact throughout history has always been limited. Great scholars such as Imam Rabbani have written that it might have been inconvenient to tell this fact to the masses and that most people would not be able to grasp it.

In the age in which we live, this fact has been made empirical by the body of evidence put forward by science. The fact that the universe is an image is described in such a concrete, clear, and explicit way for the first time in history.

For this reason, the **21st century will be a historical-turning point** when people will generally comprehend the divine realities and be led in crowds to God, the only Absolute Being. In the 21st century, it is the materialistic creeds of the 19th century that will be relegated to the trash-heaps of history, God's existence and creation will be grasped, such facts as spacelessness and timelessness will be understood, humanity will break free of the centuries-old veils, deceits and superstitions enshrouding them.

It is not possible for this unavoidable course to be impeded by any image.

CHAPTER 19

Relativity of Time and the Reality of Fate

Everything related above demonstrates that a "three-dimensional space" does not exist in reality, that it is a prejudice completely inspired by perceptions and that one leads one's whole life in "spacelessness". To assert the contrary would be to hold a superstitious belief removed from reason and scientific truth, for there is no valid proof of the existence of a three-dimensional material world.

This fact refutes the primary assumption of the materialist philosophy that underlies evolutionary theory. This is the assumption that matter is absolute and eternal. The second assumption upon which the materialistic philosophy rests is the supposition that time is absolute and eternal. This is as superstitious as the first one.

The Perception Of Time

The perception we call time is, in fact, a method by which one moment is compared to another. We can explain this with an example. For instance, when a person taps an object, he hears a particular sound. When he taps the same object five minutes later, he hears another sound. The person perceives that there is an interval between the first sound and the second and he calls this interval "time". Yet at the time he hears the second sound, the first sound he heard is no more than an imagination in his mind. It is merely a bit of information in his memory. The person formulates the perception of "time" by **comparing the moment in which he lives with what he has in his memory. If this comparison is not made, there cannot be perception of time either.**

Similarly, a person makes a comparison when he sees someone entering a room through its door and sitting down in an armchair in the middle of the room. By the time this person sits in the armchair, the images related to the moments he opens the door, walks into the room, and makes his way to the armchair are compiled as bits of information in the brain. The

The perception of time comes with comparing one moment to another. For example, we think that a period of time elapses between two people holding out their hands and then shaking them.

perception of time occurs when one compares the man sitting on the armchair with those bits of information he has.

In brief, **time comes to exist as a result of the comparison made between some illusions stored in the brain.** If man did not have memory, then his brain would not be making such interpretations and therefore the perception of time would never have been formed. The reason why one determines himself to be thirty years old is only because he has accumulated information pertaining to those thirty years in his mind. If his memory did not exist, then he would not be thinking of the existence of such a preceding period of time and he would only be experiencing the single "moment" he was living in.

The Scientific Explanation Of Timelessness

Let us try to clarify the subject by quoting explanations by various scientists and scholars on the subject. Regarding the subject of time flow-

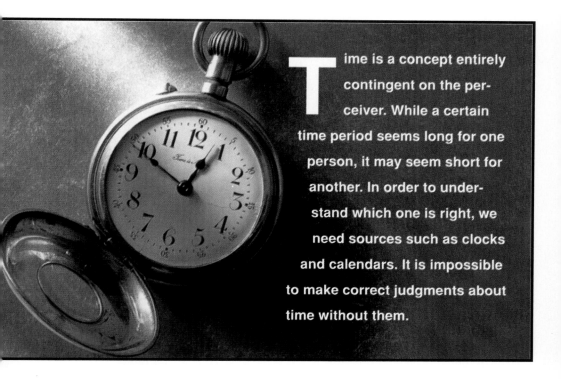

Time is a concept entirely contingent on the perceiver. While a certain time period seems long for one person, it may seem short for another. In order to understand which one is right, we need sources such as clocks and calendars. It is impossible to make correct judgments about time without them.

ing backwards, the famous intellectual and Nobel laureate professor of genetics, François Jacob, states the following in his book *Le Jeu des Possibles* (The Possible and the Actual):

> Films played backwards, make it possible for us to imagine **a world in which time flows backwards**. A world in which milk separates itself from the coffee and jumps out of the cup to reach the milk-pan; a world in which light rays are emitted from the walls to be collected in a trap (gravity center) instead of gushing out from a light source; a world in which a stone slopes to the palm of a man by the astonishing cooperation of innumerable drops of water making the stone possible to jump out of water. Yet, in such a world in which time has such opposite features, **the processes of our brain and the way our memory compiles information, would similarly be functioning backwards.** The same is true for the past and future and the world will appear to us exactly as it currently appears.[215]

Since our brain is accustomed to a certain sequence of events, the world operates not as it is related above and we assume that time always flows forward. However, this is a decision reached in the brain and therefore is completely relative. In reality, we can never know how time flows or even whether it flows or not. This is an indication of the fact that time is not an absolute fact but just a sort of perception.

The relativity of time is a fact also verified by the most important physicist of the 20th century, Albert Einstein. Lincoln Barnett, writes in his book *The Universe and Dr. Einstein*:

> Along with absolute space, Einstein discarded the concept of absolute time- of a steady, unvarying inexorable universal time flow, streaming from the infinite past to the infinite future. Much of the obscurity that has surrounded the Theory of Relativity stems from man's reluctance to recognize that sense of **time, like sense of color, is a form of perception**. Just as space is simply a possible order of material objects, so **time is simply a possible order of events**. The subjectivity of time is best explained in Einstein's own words. "The experiences of an individual" he says, "appear to us arranged in a series of events; in this series **the single events which we remember appear to be ordered according to the criterion of 'earlier' and 'later'**. There exists, therefore, for the individual, an I-time, or **subjective time**. This in itself is not measurable. I can, indeed, associate numbers with the events, in such a way that a greater number is associated with the later event than with an earlier one.[216]

Einstein himself pointed out, as quoted from Barnett's book: "space and time are forms of intuition, which **can no more be divorced from consciousness** than can our concepts of colour, shape, or size." According to the Theory of General Relativity: "**time has no independent existence apart from the order of events by which we measure it.**"[217]

Since time consists of perception, it depends entirely on the perceiver and is therefore relative.

The speed at which time flows differs according to the references we use to measure it because there is no natural clock in the human body to indicate precisely how fast time passes. As Lincoln Barnett wrote: "Just as there is no such thing as color without an eye to discern it, so an instant or an hour or a day is nothing without an event to mark it."[218]

The relativity of time is plainly experienced in dreams. Although what we see in our dream seems to last for hours, in fact, it only lasts for a few minutes, and even a few seconds.

Let us think on an example to clarify the subject further. Let us assume that we were put into a room with a single window that was specifically designed and we were kept there for a certain period of time. Let there be a clock in the room by which we can see the amount of time that has passed. During this time, we see the sun setting and rising at certain intervals from the room's window. A few days later, the answer we would give to the question about the amount of time we had spent in the room would be based both on the information we had collected by looking at the clock from time to time and on the computation we had done by referring to how many times the sun had set and risen. For example, we estimate that we had spent three days in the room. However, if the person who put us in that room comes up to us and says that we spent only two days in the room and that the sun we had been seeing from the window was falsely produced by a simulation machine and that the clock in the room was especially regulated to move slower, then the calculation we had done would bear no meaning.

This example confirms that the information we have about the rate of the passage of time is based on relative references. The relativity of the time is a scientific fact also proven by scientific methodology. Einstein's **Theory of General Relativity** maintains that the speed of time changes depending on the speed of the object and its distance from the centre of gravity. As speed increases, time is shortened, compressed; and slows down as if it comes to the point of "stopping".

Let us explain this with an example given by Einstein himself. Imagine two twins, one of whom stays on earth while the other goes travelling in space at a speed close to the speed of light. When he comes back, the traveller will see that his brother has grown much older than he has. The reason is that time flows much slower for the person who travels at speeds near the speed of light. If the same example were applied to a space-travelling father and his son staying back on earth, it would look like that: If the father was 27 years old when he set out and his son was 3, the father will, when he comes back to the earth 30 years later (earth time), be only 30, whereas the son will be 33 years old.[219]

It should be pointed out that this relativity of time is caused not by the slowing down or running fast of clocks or the slow running of a mechanical spring. It is rather the result of the differentiated operation periods of the entire material system which goes as deep as sub-atomic particles. In other words, the shortening of time is not like acting in a slow-motion picture for the person experiencing it. In such a setting where time shortens, one's heartbeats, cell replications, and brain functions, and so on all operate slower than those of the slower-moving person on Earth. The person goes on with his daily life and does not notice the shortening of time at all. Indeed the shortening does not even become apparent until the comparison is made.

Relativity In The Qur'an

The conclusion to which we are led by the findings of modern science is that **time is not an absolute fact as supposed by materialists, but only a relative perception**. What is more interesting is that this fact, undiscovered until the 20th century by science, was imparted to mankind in the Qur'an 14 centuries ago. There are various references in the Qur'an to the relativity of time.

It is possible to see the scientifically-proven fact that time is a psychological perception dependent on events, setting, and conditions in many verses of the Qur'an. For instance, the entire life of a person is a very short time as we are informed by the Qur'an;

> On the Day when He will call you, and you will answer (His Call) with (words of) His Praise and Obedience, and you will think that you have stayed (in this world) **but a little while!** (Surat al-Isra, 52)

> And on the Day when He shall gather them together, (it will seem to them) as if they had not tarried (on earth) **longer than an hour of a day:** they will recognise each other. (Surah Yunus, 45)

In some verses, it is indicated that people perceive time differently and that sometimes people can perceive a very short period of time as a very lengthy one. The following conversation of people held during their judgement in the Hereafter is a good example of this:

> He will say: "What number of years did you stay on earth?" They will say: "We stayed **a day or part of a day:** but ask those who keep account." He will say: "You stayed not but a little, if you had only known!" (Surat al-Mumenoon, 112-114)

In some other verses it is stated that time may flow at different paces in different settings:

> Yet they ask you to hasten on the Punishment! But God will not fail in His Promise. Verily **a Day in the sight of your Lord is like a thousand years of your reckoning.** (Surat al-Hajj, 47)

The angels and the spirit ascend unto him in **a day the measure whereof is (as) fifty thousand years.** (Surat al-Maarij, 4)

These verses are all manifest expressions of the relativity of time. The fact that this result only recently understood by science in the 20th century was communicated to man 1,400 years ago by the Qur'an is an indication of the revelation of the Qur'an by God, Who encompasses the whole time and space.

The narration in many other verses of the Qur'an reveals that time is a perception. This is particularly evident in the stories. For instance, God has kept the Companions of the Cave, a believing group mentioned in the Qur'an, in a deep sleep for more than three centuries. When they were awoken, these people thought that they had stayed in that state but a little while, and could not figure out how long they slept:

> Then We draw (a veil) over their ears, for a number of years, in the Cave, (so that they heard not). Then We raised them up that We might know which of the two parties would best calculate the time that they had tarried. (Surat al-Kahf, 11-12)

> Such (being their state), we raised them up (from sleep), that they might question each other. Said one of them, "How long have you stayed (here)?" They said, "We have stayed (perhaps) a day, or part of a day." (At length) they (all) said, "God (alone) knows best how long you have stayed here... (Surat al-Kahf, 19)

The situation told in the below verse is also evidence that time is in truth a psychological perception.

> Or (take) the similitude of one who passed by a hamlet, all in ruins to its roofs. He said: "Oh! how shall God bring it (ever) to life, after (this) its death?" but God caused him to die for a hundred years, then raised him up (again). He said: "How long did you tarry (thus)?" He said: (Perhaps) a day or part of a day." He said: "Nay, you have tarried thus a hundred years; but look at your food and your drink; they show no signs of age; and look at your donkey: And that We may make of you a sign unto the people, Look further at the bones, how We bring them together and clothe them with flesh." When this was shown clearly to him, he said: "I know that God has power over all things." (Surat al-Baqara, 259)

The above verse clearly emphasises that God Who created time is unbound by it. Man, on the other hand, is bound by time that God ordains. As in the verse, man is even incapable of knowing how long he stayed in his sleep. In such a state, to assert that time is absolute (just like the materialists do in their distorted mentality), would be very unreasonable.

Destiny

This relativity of time clears up a very important matter. The relativity is so variable that a period of time appearing billions of years' duration to us, may last only a second in another dimension. Moreover, an enormous period of time extending from the world's beginning to its end may not even last a second but just an instant in another dimension.

This is the very essence of the concept of destiny- a concept that is not well understood by most people, especially materialists, who deny it completely. Destiny is God's perfect knowledge of all events past or future. A majority of people question how God can already know events that have not yet been experienced and this leads them to fail in understanding the authenticity of destiny. However, "events not yet experienced" are not yet experienced only **for us**. God is not bound by time or space for He Himself has created them. For this reason, **the past, the future, and the present are all the same to God; for Him, everything has already taken place and finished**.

Lincoln Barnett explains how the Theory of General Relativity leads to this fact in *The Universe and Dr. Einstein*: According to Barnett, the universe can be "**encompassed in its entire majesty only by a cosmic intellect**".[220] The will that Barnett calls "the cosmic intellect" is the **wisdom and knowledge of God, Who prevails over the entire universe**. Just as we easily see a ruler's beginning, middle, and end, and all the units in between as a whole, God knows the time we are subjected to like a single moment right from its beginning to the end. People experience incidents only when their time comes and they witness the fate God has created for them.

It is also important to draw attention to the shallowness of the distorted understanding of destiny prevalent in society. This distorted conviction of fate holds a superstitious belief that God has determined a "destiny" for every man but that these destinies can sometimes be changed by people. For instance, for a patient who returns from death's door people make superficial statements like "He defeated his destiny". Yet, no one is able to change his destiny. The person who turns from death's door does not die because he is destined not to die then. It is again the destiny of those people who deceive themselves by saying "I defeated my destiny" to say so and maintain such a mindset.

Destiny is the eternal knowledge of God and for God, Who knows time like a single moment and Who prevails over the whole time and space, everything is determined and finished in a destiny. We also understand from what is related in the Qur'an that time is one for God: some incidents that appear to happen to us in the future are related in the Qur'an in such a way that they already took place long before. For instance, the verses that describe the account that people are to give to God in the hereafter are related as events which already occurred long ago:

> And the trumpet **is blown**, and all who are in the heavens and all who are in the earth **swoon away**, save him whom God willeth. Then it **is blown** a second time, and behold them standing waiting! And the earth **shineth** with the light of her Lord, and the Book is set up, and the prophets and the witnesses **are brought**, and **it is judged** between them with truth, and they **are not wronged**... And those who disbelieve are driven unto hell in troops... And those who keep their duty to their Lord are driven unto the Garden in troops..." (Surat az-Zumar, 68-73)

Some other verses on this subject are:

And every soul **came**, along with it a driver and a witness. (Surat al-Qaf, 21)

And the heaven **is cloven asunder**, so that on that day it is **frail**. (Surat al-Haaqqa, 16)

And because they were patient and constant, He rewarded them with a Garden and (garments of) silk. Reclining in the (Garden) on raised thrones, they **saw** there neither the sun's (excessive heat) nor excessive cold. (Surat al-Insan, 12-13)

And Hell **is placed** in full view for (all) to see. (Surat an-Naziat, 36)

But on this Day the Believers **laugh** at the Unbelievers (Surat al-Mutaffifin, 34)

And the Sinful **saw** the fire and **apprehended** that they have to fall therein: no means **did they find** to turn away therefrom. (Surat al-Kahf, 53)

As may be seen, occurrences that are going to take place after our death (from our point of view) are related as already experienced and past events in the Qur'an. God is not bound by the relative time frame that we are confined in. God has willed these things in timelessness: people have already performed them and all these events have been lived through and ended. It is imparted in the verse below that every event, be it big or small, is within the knowledge of God and recorded in a book:

> In whatever business thou may be, and whatever portion you may be reciting from the Qur'an, and whatever deed you (mankind) may be doing, We are witnesses thereof when you are deeply engrossed therein. Nor is hidden from your Lord (so much as) the weight of an atom on the earth or in heaven. And not the least and not the greatest of these things but are recorded in a clear record. (Surah Jonah, 61)

The Worry Of The Materialists

The issues discussed in this chapter, namely the truth underlying matter, timelessness, and spacelessness, are indeed extremely clear. As expressed before, these are absolutely not any sort of a philosophy or a way of thought, but crystal-clear truths impossible to deny. In addition to its being a technical reality, the rational and logical evidence also admits no other alternatives on this issue: the universe is an illusory entirety with all the matter composing it and all the people living on it. It is a collection of perceptions.

Materialists have a hard time in understanding this issue. For instance, if we return to Politzer's bus example: although Politzer technically knew that he could not step out of his perceptions he could only admit it for certain cases. That is, for Politzer, events take place in the brain until the bus crash, but as soon as the bus crash takes place, things go out of the brain and gain a physical reality. The logical defect at this point is very clear: Politzer has made the same mistake as the materialist philosopher Johnson who said "I hit the stone, my foot hurts, therefore it exists" and could not understand that the shock felt after bus impact was in fact a mere perception as well.

The subliminal reason why materialists cannot comprehend this subject is their fearing the fact they will face when they comprehend it. Lincoln Barnett informs us that this subject was "discerned" by some scientists:

> Along with philosophers' reduction of all objective reality to a shadow-world of perceptions, scientists have become aware of the alarming limitations of man's senses.[221]

Any reference made to the fact that matter and time is a perception arouses great fear in a materialist, because these are the only notions he relies on as absolute beings. He, in a sense, takes these as idols to worship; because he thinks that he is created by matter and time (through evolution).

When he feels that the universe he thinks he is living in, the world, his own body, other people, other materialist philosophers whose ideas he is influenced by, and in short, everything, is a perception, he feels overwhelmed by the horror of it all. Everything he depends on, believes in, and take recourse to vanishes suddenly. He feels the despair which he, essentially, will experience on Judgment Day in its real sense as described in the verse "That Day shall they (openly) show (their) submission to God; and all their inventions shall leave them in the lurch." (Surat an-Nahl, 87)

From then on, this materialist tries to convince himself of the reality of matter, and makes up "evidence" for this end; hits his fist on the wall, kicks stones, shouts, yells, but can never escape from the reality.

Just as they want to dismiss this reality from their minds, they also want other people to discard it. They are also aware that if the true nature of matter is known by people in general, the primitiveness of their own philosophy and the ignorance of their worldview will be bared for all to see, and there will be no ground left on which they can rationalise their views. These fears are the reason why they are so disturbed of the fact related here.

God states that the fears of the unbelievers will be intensified in the hereafter. On Judgement Day, they will be addressed thus:

> One day shall We gather them all together: We shall say to those who ascribed partners (to Us): **"Where are the partners whom you (invented and) talked about?"** (Surat al-Anaam, 22)

After that, unbelievers will bear witness to their possessions, children and close circle whom they had assumed to be real and ascribed as partners to God leaving them and vanishing. God stated this fact in the verse **"Behold! how they lie against their own souls! But the (lie) which they invented will leave them in the lurch"** (Surat al-Anaam, 24).

The Gain Of Believers

While the fact that matter and time is a perception alarms materialists, just the opposite holds true for true believers. People of faith become very glad when they have perceived the secret behind matter because this reality is the key to all questions. With this key, all secrets are unlocked.

One comes to easily understand many issues that he previously had difficulty in understanding.

As said before, the questions of death, paradise, hell, the hereafter, changing dimensions, and important questions such as "Where is God?", "What was before God?", "Who created God?", "How long will the life in cemetery last?", "Where are heaven and hell?", and "Where do heaven and hell currently exist?" will be easily answered. It will be understood with what kind of a system God created the entire universe from nothingness. So much so that, with this secret, **the questions of "when", and "where" become meaningless** because there will be no time and no place left. When spacelessness is comprehended, it will be understood that hell, heaven and earth are all actually in **the same place**. If timelessness is understood, it will be understood that everything takes place at **a single moment**: nothing is waited for and time does not go by, because everything has already happened and finished.

When this secret is delved into, **the world becomes like heaven for a believer.** All distressful material worries, anxieties, and fears vanish. The person grasps that the entire universe has a single Sovereign, that He changes the entire physical world as He pleases and that all he has to do is to turn unto Him. He then submits himself entirely to God "**to be devoted to His service**". (Surat Aal-e Imran, 35)

To comprehend this secret is the greatest gain in the world.

With this secret, another very important reality mentioned in the Qur'an is unveiled: the fact that "**God is nearer to man than his jugular vein**" (Surah Qaf, 16). As everybody knows, the jugular vein is inside the body. What could be nearer to a person than his inside? This situation can be easily explained by the reality of spacelessness. This verse can also be much better comprehended by understanding this secret.

This is the plain truth. It should be well established that there is no other helper and provider for man than God. **There is nothing but God;** He is the only absolute being Whom one can seek refuge in, appeal for help, and count on for reward.

Wherever we turn, there is the countenance of God.

CHAPTER 20

SRF Conferences: Activities for Informing the Public About Evolution

Evolution propaganda, which has gained acceleration lately, is a serious threat to national beliefs and moral values. The Science Research Foundation, which is quite aware of this fact, has undertaken the duty of informing Turkish public about the scientific truth of the matter.

First Conference-Istanbul

The first of the series of international conferences organised by Science Research Foundation (SRF) took place in 1998. Entitled "The Collapse of the Theory of Evolution: The Fact of Creation", it was held in Istanbul on April 4, 1998. The conference, which was a great success, was attended by recognised experts from around the world and provided a platform on which the theory of evolution was for the first time questioned and refuted scientifically in Turkey. People from all segments of Turkish society attended the conference, which drew a great deal of attention. Those who could not find place in the hall followed the conference live from the closed-circuit television system outside.

The conference included famous speakers from Turkey and from abroad. Following the speeches of SRF members, which revealed the ulterior ideological motives underlying the theory of evolution, a video documentary prepared by SRF was presented.

Dr Duane Gish and Dr Kenneth Cumming, two world-renowned scientists from the Institute for Creation Research in the USA are authorities on biochemistry and paleontology. They demonstrated with substantial proof that the theory of evolution has no validity whatsoever. During the conference, one of the most esteemed Turkish scientists today, Dr Cevat Babuna illustrated the miracles in each phase of a human being's creation with a slide show that shook the "coincidence hypothesis" of evolution to its roots.

Second Conference-Istanbul

The second international conference in the same series was held three months after the first on July 5, 1998 in Cemal Resit Rey Conference Hall again in Istanbul. The speakers-six Americans and one Turk-gave talks demonstrating how Darwinism had been invalidated by modern science. Cemal Resit Rey Conference Hall, with a seating capacity of a thousand, was filled to overflowing by an audience of rapt listeners.

The speakers and their subjects at this conference are summarised below.

Professor Michael P. Girouard: In his speech, "Is it Possible for Life to Emerge by Coincidences?", Michael Girouard, a professor of biology at Southern Louisiana University, explained through various examples the complexity of proteins, the basic units of life, and concluded that they could only have come into existence as a result of skilled design.

Dr Edward Boudreaux: In his speech, "The Design in Chemistry", Edward Boudreaux, a professor of chemistry at the University of New Orleans, noted that some chemical elements must have been deliberately arranged by creation in order for life to exist.

Professor Carl Fliermans: A widely-known scientist in the USA and a microbiology professor at Indiana University conducting a research on "the neutralisation of chemical wastes by bacteria" supported by the US Department of Defence, Carl Fliermans refuted evolutionist claims at the microbiological level.

Professor Edip Keha: A professor of biochemistry, Edip Keha, was the only Turkish speaker of the conference. He presented basic information on the cell and stressed through evidence that the cell could only have come into being as a result of conscious design.

Professor David Menton: A professor of anatomy at Washington University, David Menton, in a speech that was accompanied by a very interesting computer display, examined the differences between the anatomies of the feathers of birds and the scales of reptiles, thus proving the invalidity of the hypothesis that birds evolved from reptiles.

Professor Duane Gish: Famous evolutionist expert Professor Gish, in his speech entitled "The Origin of Man", refuted the thesis of man's evolution from apes.

ICR President Professor John Morris: Professor Morris, the president of the Institute for Creation Research and a famous geologist, gave a speech on the ideological and philosophical commitments lying behind evolution. He further explained that this theory has been turned into a dogma and that its defenders believe in Darwinism with a religious fervour.

Having listened to all these speeches, the audience witnessed that evolution is a dogmatic belief that is invalidated by science in all aspects. In addition, the poster exhibition entitled "The Collapse of the Theory of Evolution: The Fact of Cre-

World-renowned evolution expert Dr Duane Gish, receiving his SRF plaque from Dr Nevzat Yalcintas, A member of Turkish Parliament.

PROF. CARL FLIERMANS: "Modern biochemistry proves that organisms are marvelously designed and this fact alone proves the existance of the Creator."

PROF. DUANE GISH: "The fossil record refutes the evolutionary theory and it demonstrates that species appeared on Earth fully formed and well designed. This is a concrete evidence for that they were created by God."

PROF. DAVID MENTON: "I am examining the anatomical features of living things for 30 years. What I saw has always been the evidence of God's creation."

PROF. EDWARD BOUDREAUX: "The world we live in, and its natural laws are very precisely set up by the Creator for the benefit of us, humans."

ation" organised by the Science Research Foundation and displayed in the lobby of CRR Conference Hall attracted considerable interest. The exhibition consisted of 35 posters, each highlighting either a basic claim of evolution or a creation evidence.

Third Conference-Ankara

The third international conference of the series was held on July 12, 1998 at the Sheraton Hotel in Ankara. Participants in the conference-three Americans and one Turk-put forward explicit and substantial evidence that Darwinism has been invalidated by modern science.

Although the conference hall at the Ankara Sheraton Hotel was designed to hold an audience of about a thousand, the number of attendees at the conference exceeded 2,500. Screens were set up outside the conference hall for those who could not find place inside. The poster exhibition entitled "The Collapse of the Theory of Evolution: The Fact of Creation" held next to the conference hall also attracted considerable attention. At

Scenes from National Conferences of SRF

Ankara

Sanliurfa

Izmir

Balikesir

Samsun

Kayseri

Giresun

Bursa

the end of the conference, the speakers received a standing ovation, which proved how much the public craved enlightenment on the scientific realities regarding the evolution deceit and the fact of creation.

Following the success of these international conferences, the Science Research Foundation began organising similar conferences all over Turkey. Between August 98 and end 2002 alone, over 500 conferences were held in Turkey's all of 80 cities and towns. SRF continues to conduct its conferences in different parts of the country. SRF has also held conferences in England, Holland, Brunei, Malaysia, Indonesia, Singapore, Azerbaijan, the United States and Canada.

*Glory be to You!
We have no knowledge except what
You have taught us. You are
the All-Knowing, the All-Wise.
(Surat al-Baqara, 32)*

NOTES

1. Cliff, Conner, "Evolution vs. Creationism: In Defense of Scientific Thinking", *International Socialist Review* (Monthly Magazine Supplement to the Militant), November 1980.
2. Ali Demirsoy, *Kalıtım ve Evrim (Inheritance and Evolution)*, Ankara: Meteksan Publishing Co., 1984, p. 61.
3. Michael J. Behe, *Darwin's Black Box*, New York: Free Press, 1996, pp. 232-233.
4. Richard Dawkins, *The Blind Watchmaker*, London: W. W. Norton, 1986, p. 159.
5. Jonathan Wells, *Icons of Evolution: Science or Myth? Why Much of What We Teach About Evolution is Wrong*, Regnery Publishing, 2000, pp. 235-236
6. Dan Graves, *Science of Faith: Forty-Eight Biographies of Historic Scientists and Their Christian Faith*, Grand Rapids, MI, Kregel Resources.
7. Science, *Philosophy, And Religion: A Symposium*, 1941, CH.13.
8. Max Planck, *Where is Science Going?*, www.websophia.com/aphorisms/science.html.
9. H. S. Lipson, "A Physicist's View of Darwin's Theory", *Evolution Trends in Plants*, Vol 2, No. 1, 1988, p. 6.
10. Although Darwin came up with the claim that his theory was totally independent from that of Lamarck's, he gradually started to rely on Lamarck's assertions. Especially the 6th and the last edition of The Origin of Species is full of examples of Lamarck's "inheritance of acquired traits". See Benjamin Farrington, *What Darwin Really Said*, New York: Schocken Books, 1966, p. 64.
11. Michael Ruse, "Nonliteralist Antievolution", AAAS Symposium: "The New Antievolutionism," February 13, 1993, Boston, MA
12. Steven M. Stanley, *Macroevolution: Pattern and Process*, San Francisco: W. H. Freeman and Co. 1979, pp. 35, 159.
13. Colin Patterson, *"Cladistics"*, Interview with Brian Leek, Peter Franz, March 4, 1982, BBC.
14. Jonathan Wells, *Icons of Evolution: Science or Myth? Why Much of What We Teach About Evolution is Wrong*, Regnery Publishing, 2000, p. 141-151
15. Jerry Coyne, "Not Black and White", a review of Michael Majerus's Melanism: Evolution in Action, *Nature*, 396 (1988), p. 35-36
16. Stephen Jay Gould, "The Return of Hopeful Monsters", *Natural History*, Vol 86, July-August 1977, p. 28.
17. Charles Darwin, *The Origin of Species: A Facsimile of the First Edition*, Harvard University Press, 1964, p. 189.
18. Ibid, p. 177.
19. B. G. Ranganathan, *Origins?*, Pennsylvania: The Banner Of Truth Trust, 1988.
20. Warren Weaver, "Genetic Effects of Atomic Radiation", *Science*, Vol 123, June 29, 1956, p. 1159.
21. Gordon R. Taylor, *The Great Evolution Mystery*, New York: Harper & Row, 1983, p. 48.
22. Michael Pitman, *Adam and Evolution*, London: River Publishing, 1984, p. 70.
23. Charles Darwin, *The Origin of Species: A Facsimile of the First Edition*, Harvard University Press, 1964, p. 179.
24. Ibid, pp. 172, 280.

25 Derek V. Ager, "The Nature of the Fossil Record", Proceedings of the British Geological Association, Vol 87, 1976, p. 133.

26 Mark Czarnecki, "The Revival of the Creationist Crusade", MacLean's, January 19, 1981, p. 56.

27 R. Wesson, Beyond Natural Selection, MIT Press, Cambridge, MA, 1991, p. 45

28 David Raup, "Conflicts Between Darwin and Paleontology", Bulletin, Field Museum of Natural History, Vol 50, January 1979, p. 24.

29 Richard Monastersky, "Mysteries of the Orient", Discover, April 1993, p. 40.

30 Richard Fortey, "The Cambrian Explosion Exploded?", Science, vol 293, No 5529, 20 July 2001, p. 438-439.

31 Ibid.

32 Richard Dawkins, The Blind Watchmaker, London: W. W. Norton 1986, p. 229.

33 Douglas J. Futuyma, Science on Trial, New York: Pantheon Books, 1983, p. 197.

34 Charles Darwin, The Origin of Species: A Facsimile of the First Edition, Harvard University Press, 1964, p. 302.

35 Stefan Bengston, Nature, Vol. 345, 1990, p. 765.

36 The New Animal Phylogeny: Reliability And Implications, Proc. of Nat. Aca. of Sci., 25 April 2000, vol 97, No 9, p. 4453-4456.

37 Ibid.

38 Gerald T. Todd, "Evolution of the Lung and the Origin of Bony Fishes: A Casual Relationship", American Zoologist, Vol 26, No. 4, 1980, p. 757.

39 R. L. Carroll, Vertebrate Paleontology and Evolution, New York: W. H. Freeman and Co. 1988, p. 4.; Robert L. Carroll, Patterns and Processes of Vertebrate Evolution, Cambridge University Press, 1997, p. 296-97

40 Edwin H. Colbert, M. Morales, Evolution of the Vertebrates, New York: John Wiley and Sons, 1991, p. 99.

41 Jean-Jacques Hublin, The Hamlyn Encyclopædia of Prehistoric Animals, New York: The Hamlyn Publishing Group Ltd., 1984, p. 120.

42 Jacques Millot, "The Coelacanth", Scientific American, Vol 193, December 1955, p. 39.

43 Bilim ve Teknik Magazine, November 1998, No: 372, p. 21.

44 Robert L. Carroll, Vertebrate Paleontology and Evolution, New York: W. H. Freeman and Co., 1988, p. 198.

45 Engin Korur, "Gözlerin ve Kanatların Sırrı" (The Mystery of the Eyes and the Wings), Bilim ve Teknik, No. 203, October 1984, p. 25.

46 Nature, Vol 382, August, 1, 1996, p. 401.

47 Carl O. Dunbar, Historical Geology, New York: John Wiley and Sons, 1961, p. 310.

48 L. D. Martin, J. D. Stewart, K. N. Whetstone, The Auk, Vol 98, 1980, p. 86.

49 Ibid, p. 86; L. D. Martin, "Origins of Higher Groups of Tetrapods", Ithaca, New York: Comstock Publishing Association, 1991, pp. 485, 540.

50 S. Tarsitano, M. K. Hecht, Zoological Journal of the Linnaean Society, Vol 69, 1985, p. 178; A. D. Walker, Geological Magazine, Vol 177, 1980, p. 595.

51 Pat Shipman, "Birds do it... Did Dinosaurs?", New Scientist, February 1, 1997, p. 31.

52 "Old Bird", Discover, March 21, 1997.

53 Ibid.

54 Pat Shipman, "Birds Do It... Did Dinosaurs?", p. 28.

55 Robert L. Carroll, *Patterns and Processes of Vertebrate Evolution*, Cambridge University Press, 1997, p. 280-81.
56 Pat Shipman, *"Birds Do It... Did Dinosaurs?"*, p. 28.
57 Ibid.
58 Roger Lewin, "Bones of Mammals, Ancestors Fleshed Out", *Science*, vol 212, June 26, 1981, p. 1492.
59 George Gaylord Simpson, *Life Before Man*, New York: Time-Life Books, 1972, p. 42.
60 R. Eric Lombard, "Review of Evolutionary Principles of the Mammalian Middle Ear, Gerald Fleischer", *Evolution*, Vol 33, December 1979, p. 1230.
61 David R. Pilbeam, "Rearranging Our Family Tree", *Nature*, June 1978, p. 40.
62 Earnest A. Hooton, *Up From The Ape*, New York: McMillan, 1931, p. 332.
63 Malcolm Muggeridge, *The End of Christendom*, Grand Rapids, Eerdmans, 1980, p. 59.
64 Stephen Jay Gould, "Smith Woodward's Folly", *New Scientist*, February 5, 1979, p. 44.
65 Kenneth Oakley, William Le Gros Clark & J. S, "Piltdown", *Meydan Larousse*, Vol 10, p. 133.
66 Stephen Jay Gould, "Smith Woodward's Folly", *New Scientist*, April 5, 1979, p. 44.
67 W. K. Gregory, "Hesperopithecus Apparently Not An Ape Nor A Man", *Science*, Vol 66, December 1927, p. 579.
68 Philips Verner Bradford, Harvey Blume, *Ota Benga: The Pygmy in The Zoo*, New York: Delta Books, 1992.
69 David Pilbeam, "Humans Lose an Early Ancestor", *Science*, April 1982, pp. 6-7.
70 C. C. Swisher III, W. J. Rink, S. C. Antón, H. P. Schwarcz, G. H. Curtis, A. Suprijo, Widiasmoro, "Latest Homo erectus of Java: Potential Contemporaneity with Homo sapiens in Southeast Asia", *Science*, Volume 274, Number 5294, Issue of 13 Dec 1996, pp. 1870-1874; also see, Jeffrey Kluger, "Not So Extinct After All: The Primitive Homo Erectus May Have Survived Long Enough To Coexist With Modern Humans, *Time*, December 23, 1996
71 Solly Zuckerman, *Beyond The Ivory Tower*, New York: Toplinger Publications, 1970, pp. 75-94.
72 Charles E. Oxnard, "The Place of Australopithecines in Human Evolution: Grounds for Doubt", *Nature*, Vol 258, p. 389.
73 Holly Smith, *American Journal of Physical Antropology*, Vol 94, 1994, pp. 307-325.
74 Fred Spoor, Bernard Wood, Frans Zonneveld, "Implication of Early Hominid Labryntine Morphology for Evolution of Human Bipedal Locomotion", *Nature*, vol 369, June 23, 1994, p. 645-648.
75 Tim Bromage, *New Scientist*, vol 133, 1992, p. 38-41.
76 J. E. Cronin, N. T. Boaz, C. B. Stringer, Y. Rak, "Tempo and Mode in Hominid Evolution", *Nature*, Vol 292, 1981, p. 113-122.
77 C. L. Brace, H. Nelson, N. Korn, M. L. Brace, *Atlas of Human Evolution*, 2.b. New York: Rinehart and Wilson, 1979.
78 Alan Walker, *Scientific American*, vol 239 (2), 1978, p. 54.
79 Bernard Wood, Mark Collard, "The Human Genus", *Science*, vol 284, No 5411, 2 April 1999, p. 65-71.

80 Marvin Lubenow, *Bones of Contention*, Grand Rapids, Baker, 1992, p. 83.
81 Boyce Rensberger, *The Washington Post*, November 19, 1984.
82 Ibid.
83 Richard Leakey, *The Making of Mankind*, London: Sphere Books, 1981, p. 62.
84 Marvin Lubenow, *Bones of Contention*, Grand Rapids, Baker, 1992. p. 136.
85 Pat Shipman, "Doubting Dmanisi", *American Scientist*, November- December 2000, p. 491.
86 Erik Trinkaus, "Hard Times Among the Neanderthals", *Natural History*, vol 87, December 1978, p. 10; R. L. Holloway, "The Neanderthal Brain: What Was Primitive", American Journal of Physical Anthropology Supplement, Vol 12, 1991, p. 94.
87 Alan Walker, *Science*, vol 207, 1980, p. 1103.
88 A. J. Kelso, *Physical Antropology*, 1st ed., New York: J. B. Lipincott Co., 1970, p. 221; M. D. Leakey, Olduvai Gorge, Vol 3, Cambridge: Cambridge University Press, 1971, p. 272.
89 S. J. Gould, *Natural History*, Vol 85, 1976, p. 30.
90 *Time*, November 1996.
91 L. S. B. Leakey, *The Origin of Homo Sapiens*, ed. F. Borde, Paris: UNESCO, 1972, p. 25-29; L. S. B. Leakey, *By the Evidence*, New York: Harcourt Brace Jovanovich, 1974.
92 "Is This The Face of Our Past", *Discover*, December 1997, p. 97-100.
93 A. J. Kelso, *Physical Anthropology*, 1.b., 1970, pp. 221; M. D. Leakey, Olduvai Gorge, Vol 3, Cambridge: Cambridge University Press, 1971, p. 272.
94 Donald C. Johanson & M. A. Edey, *Lucy: The Beginnings of Humankind*, New York: Simon & Schuster, 1981, p. 250.
95 *Science News*, Vol 115, 1979, p. 196-197.
96 Ian Anderson, *New Scientist*, Vol 98, 1983, p. 373.
97 Russell H. Tuttle, *Natural History*, March 1990, p. 61-64.
98 Ruth Henke, "Aufrecht aus den Baumen", *Focus*, Vol 39, 1996, p. 178.
99 Elaine Morgan, *The Scars of Evolution*, New York: Oxford University Press, 1994, p. 5.
100 Solly Zuckerman, *Beyond The Ivory Tower*, New York: Toplinger Publications, 1970, p. 19.
101 Robert Locke, "Family Fights", *Discovering Archaeology*, July/August 1999, p. 36-39.
102 Ibid.
103 Henry Gee, *In Search of Time: Beyond the Fossil Record to a New History of Life*, New York, The Free Press, 1999, p. 126-127.
104 W. R. Bird, *The Origin of Species Revisited*, Nashville: Thomas Nelson Co., 1991, pp. 298-99.
105 "Hoyle on Evolution", *Nature*, Vol 294, November 12, 1981, p. 105.
106 Ali Demirsoy, *Kalıtım ve Evrim (Inheritance and Evolution)*, Ankara: Meteksan Publishing Co., 1984, p. 64.
107 W. R. Bird, *The Origin of Species Revisited*, Nashville: Thomas Nelson Co., 1991, p. 304.
108 Ibid, p. 305.
109 J. D. Thomas, *Evolution and Faith*, Abilene, TX, ACU Press, 1988. p. 81-82.
110 Robert Shapiro, *Origins: A Sceptics Guide to the Creation of Life on Earth*, New York, Summit Books, 1986. p.127.
111 Fred Hoyle, Chandra Wickramasinghe,

Evolution from Space, New York, Simon & Schuster, 1984, p. 148.
112. Ibid, p. 130.
113. Fabbri Britannica Bilim Ansiklopedisi (Fabbri Britannica Science Encyclopaedia), vol 2, No 22, p. 519.
114. Richard B. Bliss & Gary E. Parker, Origin of Life, California: 1979, p. 14.
115. Stanley Miller, Molecular Evolution of Life: Current Status of the Prebiotic Synthesis of Small Molecules, 1986, p. 7.
116. Kevin Mc Kean, Bilim ve Teknik, No 189, p. 7.
117. J. P. Ferris, C. T. Chen, "Photochemistry of Methane, Nitrogen, and Water Mixture As a Model for the Atmosphere of the Primitive Earth", Journal of American Chemical Society, vol 97:11, 1975, p. 2964.
118. "New Evidence on Evolution of Early Atmosphere and Life", Bulletin of the American Meteorological Society, vol 63, November 1982, p. 1328-1330.
119. Richard B. Bliss & Gary E. Parker, Origin of Life, California, 1979, p. 25.
120. W. R. Bird, The Origin of Species Revisited, Nashville: Thomas Nelson Co., 1991, p. 325.
121. Richard B. Bliss & Gary E. Parker, Origin of Life, California: 1979, p. 25.
122. Ibid.
123. S. W. Fox, K. Harada, G. Kramptiz, G. Mueller, "Chemical Origin of Cells", Chemical Engineering News, June 22, 1970, p. 80.
124. Frank B. Salisbury, "Doubts about the Modern Synthetic Theory of Evolution", American Biology Teacher, September 1971, p. 336.
125. Paul Auger, De La Physique Theorique a la Biologie, 1970, p. 118.
126. Francis Crick, Life Itself: It's Origin and Nature, New York, Simon & Schuster, 1981, p. 88.
127. Ali Demirsoy, Kalıtım ve Evrim (Inheritance and Evolution), Ankara: Meteksan Publishing Co., 1984, p. 39.
128. Homer Jacobson, "Information, Reproduction and the Origin of Life", American Scientist, January 1955, p.121.
129. Reinhard Junker & Siegfried Scherer, "Entstehung und Geschichte der Lebewesen", Weyel, 1986, p. 89.
130. Michael Denton, Evolution: A Theory in Crisis, London: Burnett Books, 1985, p. 351.
131. John Horgan, "In the Beginning", Scientific American, vol. 264, February 1991, p. 119.
132. G.F. Joyce, L. E. Orgel, "Prospects for Understanding the Origin of the RNA World", In the RNA World, New York: Cold Spring Harbor Laboratory Press, 1993, p. 13.
133. Jacques Monod, Chance and Necessity, New York: 1971, p.143.
134. Leslie E. Orgel, "The Origin of Life on the Earth", Scientific American, October 1994, vol. 271, p. 78.
135. Gordon C. Mills, Dean Kenyon, "The RNA World: A Critique", Origins & Design, 17:1, 1996
136. Brig Klyce, The RNA World, http://www.panspermia.org/rnaworld.htm
137. Chandra Wickramasinghe, Interview in London Daily Express, August 14, 1981.
138. Jeremy Rifkin, Entropy: A New World View, New York, Viking Press, 1980, p.6
139. J. H. Rush, The Dawn of Life, New York, Signet, 1962, p 35

140 Roger Lewin, "A Downward Slope to Greater Diversity", *Science*, vol. 217, 24.9.1982, p. 1239

141 George P. Stravropoulos, "The Frontiers and Limits of Science", *American Scientist*, vol. 65, November-December 1977, p.674

142 Jeremy Rifkin, *Entropy: A New World View*, p.55

143 For further info, see: Stephen C. Meyer, "The Origin of Life and the Death of Materialism", *The Intercollegiate Review*, 32, No. 2, Spring 1996

144 Charles B. Thaxton, Walter L. Bradley & Roger L. Olsen, *The Mystery of Life's Origin: Reassessing Current Theories*, 4. edition, Dallas, 1992. chapter 9, p. 134

145 Ilya Prigogine, Isabelle Stengers, *Order Out of Chaos*, New York, Bantam Books, 1984, p. 175

146 Robert Shapiro, *Origins: A Sceptics Guide to the Creation of Life on Earth*, Summit Books, New York: 1986, s. 207

147 Pierre-P Grassé, *Evolution of Living Organisms*, New York: Academic Press, 1977, p. 103.

148 Ibid, p. 107.

149 Norman Macbeth, *Darwin Retried: An Appeal to Reason*, Boston: Gambit, 1971, p. 101.

150 Malcolm Muggeridge, *The End of Christendom*, Grand Rapids: Eerdmans, 1980, sp. 43.

151 Loren C. Eiseley, *The Immense Journey*, Vintage Books, 1958, p. 186.

152 Charles Darwin, *The Origin of Species: A Facsimile of the First Edition*, Harvard University Press, 1964, p. 184.

153 Norman Macbeth, *Darwin Retried: An Appeal to Reason*, Harvard Common Press, New York: 1971, p. 33.

154 Ibid, p. 36.

155 Loren Eiseley, *The Immense Journey*, Vintage Books, 1958. p. 227.

156 Dr. Lee Spetner, "Lee Spetner/Edward Max Dialogue: Continuing an exchange with Dr. Edward E. Max", 2001, http://www.trueorigin.org/spetner2.ap

157 Ibid.

158 Ibid.

159 Francisco J. Ayala, "The Mechanisms of Evolution", *Scientific American*, Vol. 239, September 1978, p. 64.

160 Dr. Lee Spetner, "Lee Spetner/Edward Max Dialogue: Continuing an exchange with Dr. Edward E. Max", 2001, http://www.trueorigin.org/spetner2.ap

161 S. R. Scadding, "Do 'Vestigial Organs' Provide Evidence for Evolution?", *Evolutionary Theory*, Vol 5, May 1981, p. 173.

162 *The Merck Manual of Medical Information*, Home edition, New Jersey: Merck & Co., Inc. The Merck Publishing Group, Rahway, 1997.

163 H. Enoch, *Creation and Evolution*, New York: 1966, p. 18-19.

164 Frank Salisbury, "Doubts About the Modern Synthetic Theory of Evolution", *American Biology Teacher*, September 1971, p. 338.

165 Dean Kenyon & Percival Davis, *Of Pandas and People: The Central Question of Biological Origins*, (Dallas: Haughton Publishing, 1993), p. 33

166 Michael Denton, *Evolution: A Theory in Crisis*, London, Burnett Books, 1985, p. 145.

167 William Fix, *The Bone Peddlers: Selling Evolution* (New York: Macmillan Publishing Co., 1984), p. 189

168 W. R. Bird, *The Origin of Species Revisited*, Thomas Nelson Co., Nashville: 1991, pp. 98-99; Percival Davis, Dean Kenyon, *Of Pandas and People*, Haughton Publishing Co., 1990, pp. 35-38.

169 W. R. Bird, *The Origin of Species Revisited*, pp. 98-99, 199-202.

170 Michael Denton, *Evolution: A Theory in Crisis*, London: Burnett Books, 1985, pp. 290-91.

171 Hervé Philippe and Patrick Forterre, "The Rooting of the Universal Tree of Life is Not Reliable", *Journal of Molecular Evolution*, vol 49, 1999, p. 510

172 James Lake, Ravi Jain ve Maria Rivera, "Mix and Match in the Tree of Life", *Science*, vol. 283, 1999, p. 2027

173 Carl Woese, "The Universel Ancestor", *Proceedings of the National Academy of Sciences*, USA, 95, (1998) p. 6854

174 Ibid.

175 Jonathan Wells, *Icons of Evolution*, Regnery Publishing, 2000, p. 51

176 G. G. Simpson, W. Beck, *An Introduction to Biology*, New York, Harcourt Brace and World, 1965, p. 241.

177 Keith S. Thompson, "Ontogeny and Phylogeny Recapitulated", *American Scientist*, Vol 76, May/June 1988, p. 273.

178 Francis Hitching, *The Neck of the Giraffe: Where Darwin Went Wrong*, New York: Ticknor and Fields 1982, p. 204.

179 Richard Lewontin, "The Demon-Haunted World", *The New York Review of Books*, January 9, 1997, p. 28.

180 Robert Shapiro, *Origins: A Sceptics Guide to the Creation of Life on Earth*, Summit Books, New York: 1986, p. 207.

181 Hoimar Von Dithfurt, *Im Anfang War Der Wasserstoff (Secret Night of the Dinosaurs)*, Vol 2, p. 64.

182 Ali Demirsoy, *Kalıtım ve Evrim (Inheritance and Evolution)*, Ankara: Meteksan Publishing Co., 1984, p. 61.

183 Ibid, p. 61.

184 Ibid, p. 94.

185 Bilim ve Teknik, July 1989, Vol. 22, No.260, p.59

186 Grzimeks Tierleben Vögel 3, Deutscher Taschen Buch Verlag, Oktober 1993, p.92

187 David Attenborough, *Life On Earth: A Natural History*, Collins British Broadcasting Corporation, June 1979, p.236

188 David Attenborough, *Life On Earth: A Natural History*, Collins British Broadcasting Corporation, June 1979, p.240

189 "The Structure and Properties of Spider Silk", *Endeavour*, January 1986, vol. 10, pp.37-43

190 *Görsel Bilim ve Teknik Ansiklopedisi*, pp.185-186

191 WalterMetzner, http://cnas.ucr.edu/~bio/faculty/Metzner.html

192 *National Geographic*, September 1995, p.98

193 *Bilim ve Teknik*, January 1990, pp.10-12

194 David Attenborough, *Life of Birds*, Princeton Universitye Press, Princeton-New Jersey, 1998, p.47

195 James L.Gould, Carol Grant Gould, *Life at the Edge*, W.H.Freeman and Company, 1989, pp.130-136

196 David Attenborough, *The Private Life of Plants*, Princeton Universitye Press, Princeton-New Jersey, 1995, pp.81-83

197 *Encyclopedia of Reptiles and Amphibians*, Published in the United States by Acade-

mic Press, A Division of Harcourt Brace and Company, p.35

198 Frederick Vester, *Denken, Lernen, Vergessen*, vga, 1978, p.6

199 George Politzer, *Principes Fondamentaux de Philosophie*, Editions Sociales, Paris 1954, pp.38-39-44

200 R.L.Gregory, *Eye and Brain: The Psychology of Seeing*, Oxford University Press Inc. New York, 1990, p.9

201 Lincoln Barnett, *The Universe and Dr.Einstein*, William Sloane Associate, New York, 1948, p.20

202 Orhan Hançerlioğlu, *Düşünce Tarihi (The History of Thought)*, Istanbul: Remzi Bookstore, 6.ed., September 1995, p.447

203 V.I.Lenin, *Materialism and Empiriocriticism*, Progress Publishers, Moscow, 1970, p.14

204 Bertrand Russell, *ABC of Relativity*, George Allen and Unwin, London, 1964, pp.161-162

205 R.L.Gregory, *Eye and Brain: The Psychology of Seeing*, Oxford University Press Inc. New York, 1990, p.9

206 Ken Wilber, *Holographic Paradigm and Other Paradoxes*, p.20

207 George Politzer, *Principes Fondamentaux de Philosophie*, Editions Sociales, Paris 1954, p.53

208 Orhan Hançerlioğlu, *Düşünce Tarihi (The History of Thought)*, Istanbul: Remzi Bookstore, 6.ed., September 1995, p.261

209 George Politzer, *Principes Fondamentaux de Philosophie*, Editions Sociales, Paris 1954, p.65

210 Paul Davies, *Tanrı ve Yeni Fizik, (God and The New Physics)*, translated by Murat Temelli, Im Publishing, Istanbul 1995, s.180-181

211 Rennan Pekünlü, "Aldatmacanın Evrimsizliği", (Non-Evolution of Deceit), *Bilim ve Ütopya*, December 1998 (V.I.Lenin, *Materialism and Empirio-criticism*, Progress Publishers, Moscow, 1970, pp.334-335)

212 Alaettin Şenel, "Evrim Aldatmacası mı?, Devrin Aldatmacası mı?", (Evolution Deceit or Deceit of the Epoch?), *Bilim ve Ütopya*, December 1998

213 Imam Rabbani Hz. Mektupları (Letters of Rabbani), Vol.II, 357, *Letter*, p.163

214 Imam Rabbani Hz. Mektupları (Letters of Rabbani), Vol.II, 470, *Letter*, p.1432

215 François Jacob, *Le Jeu des Possibles*, University of Washington Press, 1982, p.111

216 Lincoln Barnett, *The Universe and Dr.Einstein*, William Sloane Associate, New York, 1948, pp. 52-53

217 Lincoln Barnett, *The Universe and Dr.Einstein*, William Sloane Associate, New York, 1948, p.17

218 Lincoln Barnett, *The Universe and Dr.Einstein*, William Sloane Associate, New York, 1948, p.58.

219 Paul Strathern, *The Big Idea: Einstein and Relativity*, Arrow Books, 1997, p. 57

220 Lincoln Barnett, *The Universe and Dr.Einstein*, William Sloane Associate, New York, 1948, p.84

221 Lincoln Barnett, *The Universe and Dr.Einstein*, William Sloane Associate, New York, 1948, pp.17-18

ALSO BY HARUN YAHYA

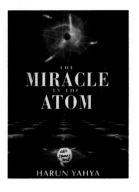

In a body that is made up of atoms, you breathe in air, eat food, and drink liquids that are all composed of atoms. In this book, the implausibility of the spontaneous formation of an atom, the building-block of everything, living or non-living, is related and the flawless nature of Allah's creation is demonstrated.
139 PAGES WITH 122 PICTURES IN COLOUR

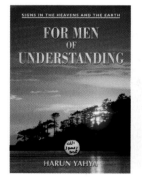

One of the purposes why the Qur'an was revealed is to summon people to think about creation and its works. When a person examines his own body or any other living thing in nature, the world or the whole universe, in it he sees a great design, art, plan and intelligence. All this is evidence proving Allah's being, unit, and eternal power.
For Men of Understanding was written to make the reader see and realise some of the evidence of creation in nature. Many living miracles are revealed in the book with hundreds of pictures and brief explanations.
288 PAGES WITH 467 PICTURES IN COLOUR

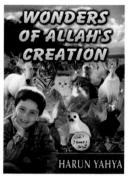

Children!
Have you ever asked yourself questions like these: How did our earth come into existence? How did the moon and sun come into being? Where were you before you were born? How did oceans, trees, animals appear on earth? How do your favourite fruits –bananas, cherries, plums– with all their bright colours and pleasant scents grow in black soil? How does a little tiny bee know how to produce delicious honey? How can it build a honeycomb with such astonishingly regular edges? Who was the first human being? Your mom gave birth to you. Yet the first human being could not have had parents. So, how did he come into existence?" In this book you will find the true answers to these questions.
144 PAGES WITH 282 PICTURES IN COLOUR

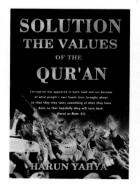

People who are oppressed, who are tortured to death, innocent babies, those who cannot afford even a loaf of bread, who must sleep in tents or even in streets in cold weather, those who are massacred just because they belong to a certain tribe, women, children, and old people who are expelled from their homes because of their religion… Eventually, there is only one solution to the injustice, chaos, terror, massacres, hunger, poverty, and oppression: the morals of the Qur'an.
208 PAGES WITH 276 PICTURES IN COLOUR

Never plead ignorance of Allah's evident existence, that everything was created by Allah, that everything you own was given to you by Allah for your subsistence, that you will not stay so long in this world, of the reality of death, that the Qur'an is the Book of truth, that you will give account for your deeds, of the voice of your conscience that always invites you to righteousness, of the existence of the hereafter and the day of account, that hell is the eternal home of severe punishment, and of the reality of fate.
112 PAGES WITH 74 PICTURES IN COLOUR

One of the major reasons why people feel a profound sense of attachment to life and cast religion aside is the assumption that life is eternal. Forgetting that death is likely to put an end to this life at any time, man simply believes that he can enjoy a perfect and happy life. Yet he evidently deceives himself. The world is a temporary place specially created by Allah to test man. That is why, it is inherently flawed and far from satisfying man's endless needs and desires. Each and every attraction existing in the world eventually wears out, becomes corrupt, decays and finally disappears. This is the never-changing reality of life.
This book explains this most important essence of life and leads man to ponder the real place to which he belongs, namely the Hereafter.
224 PAGES WITH 144 PICTURES IN COLOUR

Many societies that rebelled against the will of Allah or regarded His messengers as enemies were wiped off the face of the earth completely... All of them were destroyed–some by a volcanic eruption, some by a disastrous flood, and some by a sand storm...
Perished Nations examines these penalties as revealed in the verses of the Quran and in light of archaeological discoveries.
149 PAGES WITH 73 PICTURES IN COLOUR

Darwin said: "If it could be demonstrated that any complex organ existed, which could not possibly have been formed by numerous, successive, slight modifications, my theory would absolutely break down." When you read this book, you will see that Darwin's theory has absolutely broken down, just as he feared it would.
A thorough examination of the feathers of a bird, the sonar system of a bat or the wing structure of a fly reveal amazingly complex designs. And these designs indicate that they are created flawlessly by Allah.
208 PAGES WITH 302 PICTURES IN COLOUR

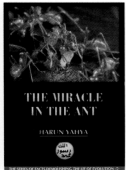

The evidence of Allah's creation is present everywhere in the universe. A person comes across many of these proofs in the course of his daily life; yet if he does not think deeply, he may wrongly consider them to be trivial details. In fact in every creature there are great mysteries to be pondered.
These millimeter-sized animals that we frequently come across but don't care much about have an excellent ability for organization and specialization that is not to be matched by any other being on earth. These aspects of ants create in one a great admiration for Allah's superior power and unmatched creation.
165 PAGES WITH 104 PICTURES IN COLOUR

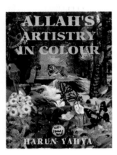

Colours, patterns, spots, even lines of each living being existing in nature have a meaning. For some species, colours serve as a communication tool; for others, they are a warning against enemies. Whatever the case, these colours are essential for the well-being of living beings. An attentive eye would immediately recognise that not only the living beings, but also everything in nature are just as they should be. Furthermore, he would realise that everything is given to the service of man: the comforting blue colour of the sky, the colourful view of flowers, the bright green trees and meadows, the moon and stars illuminating the world in pitch darkness together with innumerable beauties surrounding man...
160 PAGES WITH 215 PICTURES IN COLOUR

In the Qur'an, there is an explicit reference to the "second coming of the Jesus to the world" which is heralded in a hadith. The realisation of some information revealed in the Qur'an about Jesus can only be possible by Jesus' second coming...

Dear kids, while reading this book you will see how God has created all the creatures in the most beautiful way and how every one of them show us His endless beauty, power and knowledge. The World of Animals is also available in French.

Have you ever thought that you were non-existent before you were born and suddenly appeared on Earth? Have you ever thought that the peel of a banana, melon, watermelon or an orange each serve as a quality package preserving the fruit's odour and taste? Man is a being to which Allah has granted the faculty of thinking. Yet a majority of people fail to employ this faculty as they should... The purpose of this book is to summon people to think in the way they should and to guide them in their efforts to think.
128 PAGES WITH 137 PICTURES IN COLOUR

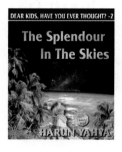

Have you ever thought about the vast dimensions of the universe we live in? As you read this book, you will see that our universe and all the living things therein are created in the most perfect way by our Creator, God. You will learn that God created the sun, the moon, our world, in short, everything in the universe so that we may live in it in the most peaceful and happy way.

Dear children, while reading this book, you will see how Allah has created all the creatures in the most beautiful way and how every one of them show us His endless beauty, power and knowledge.

MEDIA PRODUCTS BASED ON THE WORKS OF HARUN YAHYA

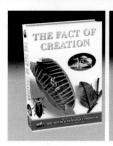  

With this CD, you will also possess a giant archive comprising the full texts of all of Harun Yahya's works, a 65-minute documentary film on how scientific discoveries confirm the miracle of the Qur'an, and 10 audio representations lasting a total of 6 hours.

HARUN YAHYA ON THE INTERNET

www.evolutiondocumentary.com
e-mail: info@evolutiondocumentary.com

www.endoftimes.net
info@endoftimes.net

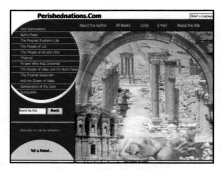

www.perishednations.com
e-mail: info@perishednations.com

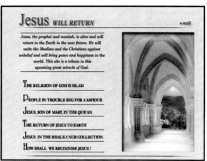

www.jesuswillreturn.com
e-mail: info@jesuswillreturn.com

www.secretbeyondmatter.com
e-mail: info@secretbeyonmatter.com

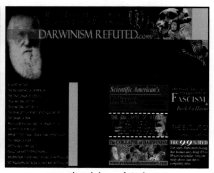

www.darwinismrefuted.com
e-mail: info@darwinismrefuted.com

www.islamdenouncesterrorism.com
e-mail: info@islamdenouncesterrorism.com

www.islamdenouncesantisemitism.com
e-mail: info@islamdenouncesantisemitism.com